Crisis in the World's Fisheries

Crisis in the World's Fisheries

People, Problems, and Policies

James R. McGoodwin

STANFORD UNIVERSITY PRESS
Stanford, California

Stanford University Press
Stanford, California

© 1990 by the Board of Trustees of the
Leland Stanford Junior University

Printed in the United States of America

CIP data appear at the end of the book

For Victor

Acknowledgments

Special thanks to Richard B. Poll-nac, Department of Sociology and Anthropology and the International Center for Marine Resource Development, University of Rhode Island, Kingston. For more than a decade he has lent encouragement and generously shared his extensive experience in the fisheries.

Also many thanks to Peter H. Fricke, National Marine Fisheries Service, National Oceanic and Atmospheric Administration, Washington, D.C.; John J. Poggie, Jr., also of the Department of Sociology and Anthropology and the International Center for Marine Resource Development; and Courtland L. Smith, Department of Anthropology, Oregon State University, Corvallis, for help and good advice on many occasions.

I am also indebted to Jan Peter Johnson and Chris Newton, both with the Fisheries Department of the Food and Agriculture Organization of the United Nations, for extending opportunities to study fishing peoples, fisheries management, and fisheries development. Also helpful have been John F. Caddy and two former members of the FAO's Fisheries Department, Menachem Ben-Yami and Ulrich Schmidt.

Many thanks also to the Woods Hole Oceanographic Institution for providing a free and stimulating environment in which to study, and especially these people for inspiration and friendship: Paul M. Fye, Peter F. Hooper, Thomas M. Leschine, Margaret Linskey, Robert W. Morse, karin negoro, Susan B. Peterson, David A. Ross, Paul R. Ryan, Leah J. Smith, Alexander Spoehr, H. Burr Steinbach, and Per Magnus Wijkman. Woods Hole, and these special people, are as snug in my memories as the catboats on Eel Pond.

These people have also helped and inspired: James M. Acheson, Raoul R. Andersen, Lee G. Anderson, Conner Bailey, David L. Bathgate, Fikret Berkes, John C. Cordell, William H. Davenport, E. Paul Durrenberger, Shepard Forman, Thomas M. Fraser, Jr., John B. Gatewood, Michael H. Glantz, John Gulland, C. P. Idyll, Svein Jentoft, R. E. Johannes, Sigurgeir Jónasson, Michael Kenny, Orvar Löfgren, John R. Maiolo, Bonnie J. McCay, Arthur F. McEvoy, Robert Lee Maril, Marc L. Miller, M. E. Moseley, Asahitaro Nishimura, Edward Norbeck, Michael K. Orbach, Evelyn W. Pinkerton, Oriol Pi-Sunyer, R. Bruce Rettig, Stuart D. Scott, M. Estellie Smith, David B. Thomson, Thorolfur Thorlindsson, Rob van Ginkel, Jojada Verrips, and William W. Warner.

At the University of Colorado, Boulder, K. Godel Gengenbach, Robert A. Hackenberg, Gordon W. Hewes, Sandra L. Karhu, and Deward E. Walker, Jr., offered encouragement and many helpful suggestions. I am also grateful to Yashka K. Hallein for excellent research assistance, Christopher J. Busick, of the Norlin Library, for help in obtaining hard-to-find publications, and Richard Y. Nishikawa for computer assistance. Finally, special thanks to William W. Carver and Karen Brown Davison of Stanford University Press for their skill, patience, and professionalism.

J.R.M.

Contents

Illustrations follow pages 64 and 122

Crisis in the World's Fisheries

Introduction

All around the world, from the coldest arctic regions to the warmest tropical seas, there is a crisis in the world's fisheries. Quite simply, there are too many people chasing too few fish, which may come as a surprise to anyone who still believes that the potential harvest of the sea is only barely being tapped.

Until fairly recently it seemed reasonable to assume that our marine resources were virtually limitless. After all, the world's total annual fish catch had steadily risen from a scant 2 million tons in 1850 to a phenomenal 55 million tons by the end of the 1960's, with little indication that any sort of upper limit was being neared.[1] However, the total catch abruptly leveled off at around 70 million tons in the early 1970's, staying around that level for the rest of the decade. During this same period the human population kept increasing, meaning that throughout the 1970's the world's per capita fish production actually declined. Correspondingly, the catch per unit of fishing effort and the catch per dollar invested in the fisheries also steadily declined, even though the number of fishing vessels grew and fishing technology became ever more effective (Brown 1978; Harris 1980: 191, 365).

What had happened? By the 1970's many coastal fish stocks had long been fished in excess of their maximum sustainable yields, and distant-water fleets that had pioneered new fishing grounds only two decades earlier were beginning to commit similar excesses.

---

[1] "Fish catch" implies the whole range of marine zoo-organisms that fishers take from the sea, including crustaceans, mollusks, and marine mammals, not merely fish. All data on fish catches are in metric tons (1 metric ton = 2,204 pounds).

Reacting to these first signs of the crisis, many fishing fleets diversified their efforts in the 1980's, ranging even more widely, using still more effective technologies, and targeting a greater variety of species than before. Some also began to seek marine species from lower trophic levels, that is, species further down the food chain. Thanks to these measures, the world catch rose again, surpassing 90 million tons by the end of the 1980's.[2] The catch per unit of fishing effort and the world's per capita fish production, however, continued to decline steadily.

Although a few fisheries experts are optimistic that the world's annual fish catch may eventually stabilize at around 100 million tons, they also stress that to reach that goal we will have to develop better means of environmental protection, as well as more effective means of fisheries management. If we do not, there is every likelihood that the world's total catch may soon begin to decline, and some stocks will eventually be wiped out permanently.

This tardy realization that the oceans and seas are producing near their upper limits is particularly alarming in view of current population projections. At this writing, the world population numbers around 5 billion—ten times as many people as just a little over 300 years ago. If current trends continue, the figure may rise to nearly 7.5 billion by the year 2000 and to between 10 and 12 billion by 2035. It is anybody's guess what level the population may ultimately reach, but even if the growth rate were to slow considerably, the near future promises to be very crowded indeed (*Economist* 1988: 8; Idyll 1978: 2–4, 7).

This is a sobering prospect, especially considering that even now between 20 and 30 percent of the world's populace suffers from severe malnutrition, and in some places from outright starvation (World Bank 1987). Certainly much of the reason for this high rate of malnutrition is the uneven distribution of food throughout the world, as well as the ecological destruction of many food-producing environments. But that is only part of the explanation. Much of the planet's potential for producing food for humans is simply being badly wasted. And not just in the realm of agriculture. In fact, nearly 30 percent of the world's total fish catch goes not for feeding people, but for reduction to fish meal and fish oils, which are used mainly in feeds for poultry and livestock. This is a great waste of the bioenergetic productivity of the sea in terms of its potential for augmenting

---

[2] Statistics for the world catch for 1850 and the 1960's are from Idyll (1978: 13); those for the 1970's and 1980's are from FAO (1970: 4, 1977: 2, 1986: 22, 1987: 20).

human food supplies, amounting to the conversion of one high-value protein source to another of similar value at the cost of considerable stored energy and protein.[3]

True, compared with terrestrially produced foods, seafoods are decidedly secondary in the global diet, contributing only around 13 percent of the total animal protein available for human consumption (Idyll 1978: 10). Nevertheless, their importance in the world's overall food supply should not be underestimated. Right now more than half of the human populace depends on fish for the majority of the animal protein it consumes, and for many of the poor in certain developing nations fish is often the only source of animal protein in their diet.

I believe that we *can* increase the sea's contribution to human food supplies, but only if we develop more effective means of managing our fisheries. Heretofore, the stress has been on conservationist/biological concerns or economic concerns, or a combination of the two. But fisheries experts are coming to pay increasing attention to fishers and fishing peoples, as they should.[4] In part this is because they recognize that fishers, particularly small-scale ones, have often confounded the best-conceived plans for managing the fisheries according to sound biological and economic principles. And it is also because of an increasing recognition that the fisheries are a human phenomenon—a recognition that, strictly speaking, there is no fishery without a human fishing effort.

In this book I want to present a fresh reconsideration of fisheries management, one emphasizing fishers and fishing peoples, which I hope will be accessible to all students of the fisheries—apprentices, able hands, and old salts alike. Beyond that, I hope to afford some interesting reading for anyone merely interested in fishers and the sea. My primary emphasis on fishers and fishing peoples is not antithetical to the earlier concerns of fisheries management; rather, it should be seen as complementary. Managers will continue to be concerned with sustaining adequate stocks while also ensuring that the greatest economic good is derived from living marine

[3] Not all marine species that are converted into fish meal are edible by humans, but fish meal itself could significantly augment food supplies in food-deficient nations, particularly if it were supplied in its highly refined form—fish protein concentrate (FPC; also called "fish flour"). But considerable efforts would be required to get consumers to accept it.

[4] Because the terms "fisherman" and "fishermen" do not take into account women who fish, I have opted instead to use "fisher" and "fishers"—terms respectably dating back to before the twelfth century.

resources. It is only that in the future, as I see it, greater emphasis must be placed on deriving the greatest *social* good from these resources.

While I am concerned with the plight of all fishers, my main focus will be on small-scale fishers, who make up the overwhelming majority of all fishers worldwide. In particular, I am interested in showing that their cultural adaptations to the special problems of their fisheries have much to tell us about how we might go about "humanizing" our fisheries policies. It is just possible that the past holds some keys to the future.

PART I

Contemporary Fishers and Fisheries

And they spoke politely about the current and the depths
they had drifted their lines at and the steady good weather
and of what they had seen.

Ernest Hemingway,
The Old Man and the Sea

Fishers Today: Between the Devil and the Deep Blue Sea

Fishing, when thought of as an occupation or a means of livelihood, evokes a variety of images in people's minds. For some it is a romantic image, such as Hemingway's old man of the sea. For others it is a small, rundown fleet of ragtag trawlers, sitting at their berths in an urban seaport. Fishing is often big business, of course, and for many urbanites big business is mainly what comes to mind—a fleet of agile tuna seiners on the hunt in tropical waters, or a huge factory ship plowing through heavy seas in a cold and remote northern ocean, burly fishermen at the stern, dressed in heavy rubberized slickers, hauling back nets full of fish, dumping the catch on deck, sorting it, and sending it down chutes to be processed, packaged, and quick-frozen below. For most people, however, fishing connotes something decidedly smaller in scale—a fleet of small work boats scattered across a bay, for instance, or several dozen canoes and outriggers pulled up on the beach next to a vacation hotel in an exotic, tropical country.

Although most people in the world's cities and towns consume seafoods at one time or another, few give more than a passing thought to the world's fishers. Indeed, many of them live inland and have little direct contact with the sea, which considerably impoverishes their understanding of what fishing peoples are all about. Thus it is no surprise that many people think of fishing only during a reverie—a reminiscence, perhaps, of a picturesque fishing fleet seen while on a summer vacation. Actually, touring urbanites seldom pay much attention to the local fishers who inhabit their playgrounds—unless, perhaps, the fishers paint their boats in vivid colors or hold quaint traditional festivals. Even in coastal resorts, where local fishers usually supply nearly all of the seafoods served in the restaurants and hotels, they often go practically unnoticed by the seasonal tourists.

Many landlubbers also assume that the lives of fishers are comparatively free and uncomplicated. Few are aware of how complex and difficult the occupation really is. Still fewer are aware of the extent to which fishers contribute to local and national economies, or the extent to which they supply high quality, protein-rich foods to a hungry world.

Before we can talk about fisheries policies, then, we need to know more about the fishers themselves, and particularly the small-scale fishers, who for historical and logistical reasons have been largely neglected in the formulation of fisheries policies. That neglect is the more astonishing considering that small-scale operators constitute fully 94 percent of the world's fishers—some 8 million people—and produce nearly half the world's fish catch designated for human consumption (Thomson 1980: 3).[1] If we include all of the ancillary workers who support these small-scale producers, and all the dependents of both, we are talking about roughly 100 million people who are associated with small-scale fishing.

Small-Scale Fishers

Let us begin by defining our terms. In the literature on fisheries and fishing peoples, as well as in the everyday parlance of national and international development agencies, scientific institutions, and so forth, small-scale fishers are identified in a variety of ways. Among development organizations, a favorite tag is "artisanal" fishers. In other circles, we find the terms "native," "coastal," "inshore," "tribal," "peasant," "traditional," or "small-scale." In essence, what all of these fishers have in common is their relatively small capital commitment.[2]

[1] Thomson provides other data comparing small- and large-scale fishers. He concludes that by nearly all measures small-scale fishers are superior to their large-scale, industrialized counterparts. Assuming Thomson's main sources were FAO statistics, his data may actually underestimate the overall number of small-scale fishers, since the FAO has been mainly concerned with impoverished "tribal," "peasant," and "artisanal" fishers in the developing nations, and may not include in its data less-impoverished ones, particularly those living in the developed nations. Ben-Yami (1980), who also presumably relied on FAO statistics, more conservatively estimates the proportion of small-scale fishers at around 90 percent, harvesting around 33 percent of the food fish consumed worldwide. Data similar to those of both Thomson and Ben-Yami can be found in Bailey (1988b) and Lawson (1984).

[2] Besançon (1965: 163, 175) makes a distinction between small- and large-scale fisheries on the basis of cultural diversity, with small-scale fisheries manifesting a great variety of cultures, and large-scale fisheries reflecting a collapsing of cultural diversity as a result of the mechanization and scientific innovation introduced by the more technologically advanced nations. Luna (n.d.: 1) objects to the term "small-scale." "'Small-scale fisheries,'" he states, "is not a concept sufficiently precise and specific . . . since the 'scale' is variable. . . . It is more acceptable to make the distinction between 'Artisanal Fisheries' and 'Industrial Fisheries.'"

The label "artisanal," favored by various international development agencies, such as the Food and Agriculture Organization of the United Nations (FAO), is quite apropos for describing fisher-artisans who fabricate much of their own gear, perhaps by weaving their own nets, fashioning fish pots or traps with palm fronds and cordage, or employing crude, homemade watercraft, such as log rafts or dugout canoes. Julio Luna (n.d.: 1) notes that the artisanal fishers' art is the skill, experience, and intuition they apply to their fishing effort. But, the term "artisanal" hardly seems apropos when applied to fishers who use small motorized watercraft with fiberglass hulls, for example, or fishing gear that is nearly all purchased or supplied from sources outside their local communities. Yet quite often, and confusingly, these people are still identified as artisanal in fisheries management and development contexts.

For some, the crucial distinction is between "inshore," "coastal," or "near-shore" fishers and "deep-sea" or "distant-water" fishers, but I do not find this distinction particularly satisfactory either, because there are some fishers with very large capital commitments who still fish quite close to shore.

Richard B. Pollnac, who has had considerable experience with a great variety of fishing peoples around the world, has convinced me that for purposes of talking about the problems of artisanal, peasant, tribal, traditional, and other such fishing peoples, the term "small-scale" is the most useful, mainly because it is the most encompassing. It circumscribes a diversity of maritime peoples around the world who share similar problems, and whose individual capital commitments and levels of production are relatively small-scale. Obviously, some definitional precision is lost through blanket employment of the term, for example when it is used to refer to both tribal and peasant fishers, who otherwise have considerably different cultures. Nevertheless, the tag seems both descriptively valid and operationally useful when discussing either of these groups in the context of fisheries management, because in this larger context the groups find themselves in a similar political position vis-à-vis the establishment of fisheries management policies, they both experience culture strain and economic

---

McCay (1981a) provides a very thoughtful examination of definitional problems. Of particular interest is her discussion of the small-scale fisher as "peasant." She questions the extent to which fishing people are like farmers and whether they face the same developmental issues. The main difference, as she sees it, is the "overfishing problem," for though the overuse of resources may be an important issue among agricultural peasants, it is usually a crucial one for fishing peoples.

and political marginalization as a result of the industrialization of fishing, and they both contribute to the resource depletions that increasingly worry fisheries experts.

Because small-scale fishing implies a small-scale capital commitment, it also usually implies small-scale power, that is, an inability to influence fish markets, little representation in the formulation and implementation of fisheries management policies, and an inability to safeguard fisheries against the environmental degradation caused by external developments. This is particularly true of small-scale fishers in the developing nations. Because of their great number and decentralization, and sometimes because of their cultural heterogeneity, it is difficult for such fishers to organize collectively for their common welfare. This often places them at a great competitive disadvantage with large-scale fishers, who form powerful organizations to secure favorable marketing arrangements for their catches and to lobby government officials.

While the majority of small-scale fishers are found in the developing nations, a considerable number can be found along the coastlines of the developed nations as well. Though the latter may employ more sophisticated types of fishing gear and measure their daily catches in tons rather than in kilograms, they may still qualify as small-scale in terms of their capital commitment. And though many of them occupy the low end of the social, economic, and political continuum, some earn incomes and enjoy standards of living that are comparable to or better than the national averages of their countries. This is rarely the case in the developing nations, however, where small-scale fishers tend to be among the poorest of the poor. Some have been traditional fishers for a long time, and because fishing is the only way of life they have known for generations they may persist in it even though it provides only a very marginal livelihood.

A growing number of fishers in these countries are also comparative newcomers to fishing. Often they are severely impoverished farmers who have lost their lands or cannot subsist on them, and, unable to find work in the growing sectors of the national economy, have migrated to the land's end. They often arrive with little more than a naïve dream of somehow extracting a living from what seems a comparatively free, commonly owned, and essentially uninhabited domain. Yet once there they may find themselves engaged in an even more desperate struggle to survive.

Small-scale fishers comprise an astounding diversity of peoples and cultures scattered around the world. In the developing areas of the Western hemisphere, we find impoverished rural Mexican *campesinos,* throwing cast nets for shrimps in a shallow lagoon; the Miskito Indians in eastern Nicaragua, who capture sea turtles for a commercial market; and Brazilian raft fishers who work their country's mangrove swamps, as well as far offshore.[3]

Many other small-scale fishers can be found along the coastlines of Africa, the Middle East, and still farther east. They include the canoe fishers along the coast of West Africa, who fish at night by lantern light; Egyptians, Sudanese, and Arabs fishing in the Red Sea from graceful dhows; and other Moslem fishers farther east, in the Persian Gulf, in the Bay of Bengal, along the Malay peninsula, and throughout the Indonesian archipelago. We also have the maritime peoples of Micronesia, Melanesia, and Polynesia, as well as the so-called "sea gypsies" living around the Philippine Islands, who spend nearly all their lives aboard junks and sampans, and the countless other maritime peoples inhabiting the shores of India and the mainland coasts of South and Southeast Asia.

An astonishingly large number of equally diverse small-scale fishers can also be found in the developed nations. In the United States, for example, we have the Eskimo seal hunters of northern Alaska, the ruggedly individualistic New England lobster fishers, the New Jersey and Chesapeake Bay watermen, sponge fishers descended from Greek immigrants in Florida, father-and-son salmon trollers in Northern California, and Vietnamese immigrants setting crab pots in the inshore bays of Texas or fishing in California's Monterey Bay.[4]

Likewise, around much of the North Atlantic rim, in such countries as Canada, Iceland, Denmark, Norway, Scotland, and Ireland, one still may find fisher-crofters, crab-pot fishers, net fishers who employ oar-powered wooden skiffs, and so forth. Small-scale fishers in other modernized Western European nations include the Portuguese whalers from Pico Island in the Azores, who still hunt whales from wooden skiffs using hand-thrown harpoons, and free-lancers of every sort plying the coasts of France, Spain, Italy, and Greece.

The list could be extended endlessly: to the more developed nations of Asia—Japan, Korea, and Taiwan—to Australia and New Zealand, to the farthest reaches of the Soviet Union. In short, small-scale fishers work in every corner of the world, from the coldest arctic region to the warmest tropical sea.

Starting with primitive fishers and working our way up through tribal,

[3] Offshore fishing from crude rafts, called *jangadas* by Brazilian fishers, is described by Forman (1967, 1970). Cordell (1973) contains an excellent review of Forman (1970).

[4] See Orbach (1983) for a discussion of the "success in failure" of immigrant Vietnamese fishers in Monterey Bay.

peasant, artisanal, and small-scale commercial fishers, we move up a developmental continuum marked by an increasing degree of capitalization, technological sophistication, and catch sizes. Somewhere further up this hypothetical continuum we encounter intermediate-scale and finally large-scale fishers. Unfortunately, the definitional boundaries between these various categories are fuzzy and we can clearly distinguish them only when they are at the opposite, extreme ends of this continuum.

Though the degree of capital commitment largely determines a fishing enterprise's place along the continuum, vessel size alone is not a reliable determinant. A high-tech tuna seiner that is over 200 feet in length and capable of carrying nearly 2,000 tons of fish in its hold, but is owned and operated by a local kinship group, is not "large-scale, industrialized fishing" on the same scale as a similar boat that is part of a fleet under the professional management of a large corporation. Similarly, a small shrimp trawler may be representative of large-scale, industrialized fishing when it is part of a corporately owned fleet organized to supply shrimp to a centralized processing and packing plant.

There is one category of small-scale fishers of great concern to fisheries managers, but with whom I will not be particularly concerned in this book: recreational fishers. This is because I want to limit my focus to people who derive their primary means of livelihood from the sea and whose primary purpose is to produce seafoods for human consumption. Certainly recreational fishers do constitute large numbers of people, but because their primary interest is, in fact, recreational, I feel their interests in the fisheries are secondary. Nevertheless, their impact on fisheries resources and national economies should not be underestimated. Frederick W. Bell (1978: 269), for example, cites some astounding statistics prepared by Nathan Associates (1974) comparing expenditures on recreational and commercial fishing in the United States. In 1970, gross expenditures on saltwater recreational fishing were an estimated $4.96 billion; by comparison, *commercial* domestic landings in 1973 had an estimated retail value of only around $3.0 billion. Since then, recreational fishing has undergone a boom, not only in the United States, but in countless coastal resort areas around the world. Thus, its impact on fish stocks poses increasingly serious problems for commercial fishers—problems that are compounded even more by recreational fishers' increasing political clout (see Stoffle et al. 1983).[5]

[5] For more recent studies of the recreational fishing industry and the problems it poses for fisheries management, see Kitner and Maiolo (1988); Stroud, ed. (1984); R. B. Thompson (1984); W. G. Gordon (1981); Dawson and Wilkins (1980); and Clepper, ed. (1979, 1981).

Modernization

Whether the poorest of the poor in a developing nation or middle-class people in a developed nation, fishers usually have high-risk jobs, low occupational mobility, uncertain incomes, and chaotic family lives, and they often find themselves caught in an increasingly competitive struggle with other fishers. Moreover, they are often held in low esteem by the non-fishing populace and occasionally even by some fisheries management professionals. But these are only the more obvious problems plaguing fishers today, particularly the small-scale ones. There are others, that trouble them just as greatly but are more subtle. Modernization processes, particularly, have taken a heavy toll.

In the fisheries, no less than in other sectors of modern economies, the process of modernization has favored forms of economic growth characterized by the accumulation of capital. This in turn has made possible rates of exploitation of natural resources such as the world has never seen before. In many fisheries, these high levels of production have brought about serious marine-resource depletions as well as other forms of environmental degradation. In human terms, the cost has been severe socioeconomic dislocations among the world's fishing peoples.

For many fishing peoples, modernization has prompted longer working hours, unemployment, greater economic risk, the depletion of vital resources, the disruption of traditional modes of subsistence, the marginalization of what had heretofore been satisfying and viable lifestyles, and the disintegration of well-established patterns of social life. Sometimes with changes in fishing technology, important cooperative activities have become unnecessary—the need to carefully hang up nets to dry, for instance, which is no longer necessary once nylon nets replace those made of cotton. Also there have been shifts to more impersonal modes of recruiting fellow workers. Old patterns of relations have given way to more individualistic and competitive patterns, with the result that the organization of many fishing communities has become more atomistic. Many formerly cooperative fishing peoples have become more guarded in their dealings with one another and have found themselves embroiled in an increasing number of conflicts with other community members.

Modernization and modern lifestyles and values have also fueled some rather crazy production, marketing, and distribution patterns in certain fisheries—for instance, in those supplying species of fish that confer social prestige on consumers. In such fisheries we often find excessively high market demands placed on scarce species while other abundant and equally nutritious stocks remain underutilized. E. N. Anderson, Jr. (1975: 40), for example, describes how pomfrets and prawns caught by Hong Kong fishers are objects of conspicuous consumption that reinforce consumers' social status, with the result that the high demand has severely reduced these stocks, while abundant stocks of red snapper, jack fish, and other excellent protein sources remain underutilized.

Modernization and technological change in the fisheries have often come about fairly rapidly, as when new competitors force local fishers to adopt new types of fishing gear, or when larger markets for production present themselves. Quite often technological changes have been stimulated and then guided by various development organizations. And sometimes the seductive allure of the new technologies themselves has prompted fishers to adopt them, especially the so-called high-tech electronics, which are feverishly promoted by national and multinational business organizations.[6]

Generally urban-based, highly capitalized, and seen by national governments as more significant and economically promising, the large-scale sector usually enjoys government support, sometimes even direct subsidization, to a degree practically unheard of in the small-scale sector of the fisheries.[7]

Still, this is not to say that modernization has always benefited large-scale fishers at the expense of small-scale ones. Indeed, because marine environments have only a finite amount of seafoods, and because modernization often fuels market demands in excess of what these marine environments can produce, both types of fishers have often been victimized by their own fishing activities during the modern era. And both sectors share the problems created by petroleum and mineral exploitation in the seas: obstructions on the seabed and in the water column, not to say pollution. As an increasing volume of polluting effluents from modern civilization finds its way into the seas, more and more fisheries have had to be closed down because their marine life has been decimated or rendered unfit for human consumption. Sometimes these catastrophes develop suddenly and precipitously over a very short period of time, as in the case of sudden well

---

[6] Alexander (1975) describes how even the best-intentioned efforts of developmentalists have done more harm than good for the small-scale fishers of Sri Lanka. Their error was trying to introduce new technologies without adequately considering the potential for disemployment, among other deleterious societal consequences. Fraser (1960) is a landmark work on this point, showing how the adoption of nylon nets and powered boats brought about strains in the rural fishing communities of south Thailand.

[7] See McGoodwin (1987), a case study of the marginalization of small-scale rural Mexican fishers following the subsidized development of the fish-export industry.

blowouts or shipping accidents. More insidiously, they often develop imperceptibly, until pollution finally and fully degrades a once-productive marine environment. These problems impinge on all fishers, but not nearly to the same extent. Because their incidence is usually greater closer to shore, where nearly all small-scale fishers are found, their relative impact is especially severe on small-scale fishers.

While the solution to pollution is prevention, such as by imposing heavy taxes or penalties on the dispersal of polluting effluents, even in conservation-minded countries like the United States decisive solutions have been slow in coming. One of the problems is that the sources of pollution are often numerous, making it difficult to identify the wrongdoers. Further, the dispersal of pollutants in the sea is quite diffuse, making it difficult to reasonably calculate the costs in terms of losses for the fisheries and for other users of water resources. Even when the sources of pollution can be pinpointed and the costs reasonably calculated, it is still usually difficult to hold the polluters responsible because they often have tremendous political and economic clout. Adding to these problems is a cacophonous assemblage of advocacy and interest groups that often helps to confuse the process of establishing sound remedial policies. Thus most fishers are rarely compensated for the reduction of economic benefits that pollution brings about in their fisheries.

Although ocean pollution is currently one of the most serious and urgent problems facing the planet, the establishment and implementation of an appropriate and workable international policy has proved to be a knotty problem indeed. Even now, over two decades after a new environmental movement began to influence pollution policy in the United States, prompting the enactment of legislation and the creation of a federal agency to safeguard the environment, ocean pollution still menaces, seemingly beyond our control.[8]

Another aspect of the modernization phenomenon that has had a negative impact on fishers, and particularly on small-scale fishers, is the rapid growth of the tourist, recreation, and leisure industries. Fishers have often

[8] Indeed, as I was completing this book, the United States experienced the worst ocean pollution accident in its history when an oil tanker struck a reef in Prince William Sound. This region has one of the most productive fisheries along the entire Alaska coast. Bell (1978: 205–38) contains an excellent discussion of the increasingly harmful impact of waterborne pollutants on the world's fisheries and gives estimates of the economic costs to some of them (only rarely borne by those responsible). Leschine (1988) discusses the immense problems surrounding the establishment and implementation of ocean pollution policies.

been displaced from traditional fishing harbors by marinas and yacht basins, and hotels have been constructed on beaches once used as important fish landing and processing sites. Tourists have also sometimes disrupted or subverted local traditions and the stability of social life in fishing communities with their more pleasure-oriented values.[9]

Probably the most threatening aspect of the burgeoning recreation and leisure industries, however, is the phenomenal growth of recreational fishing in many parts of the world. In a few fisheries recreational fishing has grown so large that its yield now surpasses that of commercial fishing. Thus recreational fishers have become an increasingly formidable lobby, with an ability to influence fisheries conservation and allocation policies. In many policy-making arenas, the recreational sector of the industry has persuasively argued that compared with the commercial sector, recreational fishing makes a greater overall economic contribution to society at large and does more to increase general well-being by providing greater personal satisfaction. As Kathi R. Kitner and John R. Maiolo (1988: 213) observe, "In the area of coastal and ocean resource policy . . . commercial and private recreational interests often are pitted against each other in heated controversy."

The problem recreational fishing poses for policymakers is also complicated by definitional problems that make it difficult to distinguish precisely between recreational and commercial fishers. Sometimes recreational fishers make such large catches that they sell them in commercial markets; conversely, commercial fishers sometimes fish recreationally or charter their boats to people who fish strictly for leisure and recreational purposes. Whatever the case, the phenomenal growth of recreational fishing is particularly threatening to small-scale fishers for whom fishing is the main means of subsistence and the primary source of self identity (see Andersen 1982; Joseph 1979; Maiolo 1981; Maiolo and Tschetter 1982; Meltzoff and Lipuma 1986a, b).

Also perplexing to many fishers has been the worldwide environmental movement, which has sometimes pitted fishers against environmentalists:

---

[9] Problems posed for fishers as a result of the development of tourism are discussed by Pi-Sunyer (1976: 60–68), who describes how fishing peoples living in a Spanish community lost prime harbor and shore space. J. C. Johnson and D. Metzger (1983: 429) discuss the situation in Southern California, where "the expressive/recreational use of [the] harbors is displacing their technical/commercial uses." The deleterious consequences of the tourism industry in rural coastal communities in developing nations is described as a "tourism-impact syndrome" in McGoodwin (1986), which generalizes from the case of a rural fishing community on the Pacific coast of Mexico.

the Newfoundland sealers and Japanese coastal fishers versus Greenpeace, for example, and the abalone fishers along the central California coast, who bitterly watched the near-ruination of their industry with the effort to protect the formerly endangered sea otter. As management regimes become increasingly more complex, and bureaucratic control over fisheries ever more top-heavy, today's fishers must similarly become more knowledgeable, sophisticated, and politically active. John R. Maiolo and Michael K. Orbach observe (1982: 7):

For fishing peoples . . . modernization is not only the process of adopting new physical materials . . . and organizational technologies in their occupation and in their lifestyles, but also the process of learning to work within the new political and administrative constraints and requirements imposed by the upward aggregation of management regimes. . . . Fishermen . . . must "modernize" in that they must acculturate themselves to the ideas and behaviors of those at all levels to which the authority and responsibility for the management of fisheries have been aggregated.

The foregoing comments notwithstanding, it would be erroneous to conceive of fishers as passive victims of the modernizing, techno-corporate state. Indeed, what is fascinating—and also tragic—about the fishing industry is that it so actively participates in its own annihilation.

Other Problems

Whether fishing was ever at one time a free and unencumbered way of making a living, it is definitely not anything close to that today for most fishers. Moreover, it promises to become even more complicated in the future. Most fishers nowadays are caught between several devils and the deep blue sea. Ashore they are bedeviled by families troubled by long absences and sporadic incomes, community members and a general public who often hold them in low esteem, and intermediaries who absorb a large share of their earnings. They often find themselves having to deal with unpredictable and fluctuating markets, production costs that rise faster than the prices paid for their catches, and the presence of many other people whom they regard at best as meddlesome and at worst as threatening. These people may include fisheries managers, who many fishers feel implement regulations in an imperious manner; misguided developmentalists, who may involve fishers in ruinous culture change; environmentalists who are critical of fishers' practices and who may threaten their economic well-being; tourists who compete for valuable shore space while sometimes offending fishers' personal values; and nowadays, coast guard and other authorities who board their boats or otherwise interrupt fishing activities in the effort to interdict illegal drug traffic.

Another problem might be termed definitional. Though many of those in the large-scale, industrialized sector are more or less full-time, year-round fishers, for whom fishing may be the sole means of making a living, most small-scale fishers are part-time or seasonal operators who have other means of livelihood as well. This fact is sometimes misunderstood by fisheries managers, especially those who attempt to draw a distinction between fishers and nonfishers for management purposes, artificially forcing that distinction by resorting to such labels as "part-time" or "full-time," and "seasonal" or "year-round." Distinguishing fishers from nonfishers on the basis of such criteria is seldom valid, because few coastal marine environments offer the possibility of daily, year-round fishing activity. Similarly, in most coastal areas, there are inevitably people who occasionally fish but who should not be occupationally identified as fishers: recreational fishers, who fish mainly for pleasure and to fill their leisure time, or people who may simply economically exploit sudden and unusual windfalls of fish.

Any ascription of an occupational identity as a "fisher" must therefore be based on the level of commitment the person feels toward the occupation and the importance of that occupation in the person's life. I believe two minimal criteria should be employed: first, self-identification as a fisher, a designation that could be corroborated by most community members; and second, the person's reliance primarily on the capture of marine organisms for his or her livelihood or standard of living.

To be sure, many of the problems faced by fishers are shared by nonfishers, but fishers are also bedeviled by specific problems they face at sea: diminishing fish stocks and a climate of ruthlessly intensifying competition; the degradation of marine environments by pollution, which often originates far beyond their sphere of control; and recreational fishers who compete for marine resources and whose economic impact may loom more significant in the minds of fisheries policymakers than the economic contributions of the local commercial fleet. Moreover, many small-scale fishers now find themselves increasingly losing competitive struggles with industrialized fishers from urban ports in their own countries, or with fishers who have come from distant lands. As Conner Bailey (1987: 173) notes, large-scale fishing vessels such as trawlers, operating in coastal waters, may "not only compete effectively against small-scale fishermen, but because of their active mode of operation they frequently damage or destroy more passive small-scale gear."

Most fishers also suffer from the difficulty of claiming ownership of living marine resources, which are still designated as common property in most fisheries around the world. This leaves local fishers particularly vulnerable to incursions by "outsiders," since property rights in fisheries are more difficult to define and assert than rights in land-based resources. Moreover, the governments of developing nations often lack sufficient power to keep other competitors—both domestic and foreign—from operating in their traditional fishing grounds. According to Kenneth Ruddle and Tomoya Akimichi (1984: 1), "uncertain, weak or contested tenurial status is one of the principal difficulties encountered by small-scale fishermen."[10]

And notwithstanding all the foregoing maladies, all fishers must still cope with the sea itself, which has always been one of the most uncertain, destructive, and dangerous working environments in the world.

It may well be that today's fishers face an even greater degree of risk and uncertainty than their predecessors did in premodern times. The maritime anthropologist M. Estellie Smith (1988) notes that while today's fishers face most of the risks and uncertainties centering on personal safety and economic concerns that fishers have always faced, they also struggle with risks and uncertainties that either are new or constitute old problems now greatly aggravated—for example, the increased social risks posed by the adoption of more modern fishing vessels, which compels them to stay at sea longer, thereby increasing their alienation and estrangement from their families and communities, and the increasing territorial risks as their fishing grounds become more crowded and competitive. They also confront greater political risks as fisheries management regimes become increasingly more complicated and change with greater rapidity.

So how do fishers cope with these formidable problems? In most cases they muddle through, striving to live as they always have, pursuing the only life they know, attempting to adapt as they can, trying to work around factors that threaten their existence, avoiding whatever seems most threatening at the moment. Moreover, in spite of the implementation and enforcement of modern management regimes in their fisheries, many still attempt to regulate fishing activities themselves.

---

[10] Two reviews published in the early 1980's, Acheson (1981b) and Emmerson (1980), both underscore the difficulties small-scale fishers face when attempting to assert their "sea tenure" rights. Fishers' sea tenure and territoriality are explored more extensively in Cordell, ed. (1989).

When we consider most fishers today, we are examining a type of existence usually characterized by hard outdoor work, relatively low incomes, rugged individualism, a high degree of risk taking, the capture of common property resources, low public esteem, and high degrees of dispersion and decentralization—all of which contribute to making fishing an exceedingly difficult mode of existence, as well as a very difficult activity to manage.

The Cultures of Fishing Peoples

A fairly new specialization, tracing its beginnings only from around the close of the Second World War, maritime anthropology focuses mainly on fishers and fishing peoples, human maritime adaptations, and such topics as early navigation and prehistoric maritime societies.[1] It is a discipline that has benefited considerably from the increasing sophistication of the ethnographic studies of the 1930's and 1940's, as well as from financial support for postwar reconstruction and development, and that gained even greater impetus—and support—after the 1950's, when new studies of fishers were prompted and in some cases mandated by important changes in national and international ocean management policies.[2]

In general, most of the studies of fishing peoples published after the war transcended the state of the art reflected by their predecessors. Earlier studies were seldom problem-focused, almost never comparative, and mainly concerned with describing exotic ways of life and primitive fishing gear. More recently, however, scholars have explored not only the fundamental characteristics of maritime cultures,[3] but also how those cultures differ from land-based cultures, the unique patterns of cultural development in

[1] Gatewood and McCay (1988: 103–4) draw an interesting distinction between "maritime anthropology" and the "anthropology of fishing." The research emphasis of the former, they state, is community studies of fishing peoples; that of the latter is mainly the activity of fishing itself. While these distinctions are undoubtedly useful in certain contexts, both emphases constitute "maritime anthropology" for me.

[2] Notable works in maritime anthropology published during the 1970's and 1980's include Acheson (1981b); Andersen, ed. (1979); Casteel and Quimby, eds. (1975); Cordell, ed. (1989); Fricke, ed. (1973); Nishimura (1973); Ruddle and Akimichi, eds. (1984); M. E. Smith, ed. (1977); and Spoehr, ed. (1980).

[3] *Cultures* is here meant to be synonymous with *sociocultures*. "Culture" has somewhat greater currency in American anthropology, while "socioculture" is preferred by British anthropologists. Whether it is more appropriate to speak of "cultural anthropology" or "sociocultural anthropology" has been the subject of debate among anthropologists, but this debate

maritime societies, relationships between culture and ecology, local systems of sea tenure and territoriality, and especially problems threatening fishing peoples' general welfare. Nearly all studies of fishing peoples in maritime anthropology address the need for more appropriate fisheries management policies.

A few anthropologists doubt whether maritime cultures are a unique genus that should be considered distinct from land-based cultures (see, e.g., Bernard 1976: 478–79). In essence, they argue that despite the extreme adaptations prompted by marine environments, all we are seeing in maritime cultures is basic human nature acted out in maritime settings. However, this theoretical debate is essentially irrelevant to most applied social scientists, who stress that the development of wiser fisheries management policies necessarily requires us to focus on peoples who derive their livelihood from the sea.

Whenever a people share a common means of making a living, honor the same or similar traditions and beliefs, are centered around a community or inhabit a region comprising a set of similar communities, and live in the same communities in which their families and other kinfolk live and work, they presumably share a common and distinct culture. Such peoples share not only the same physical space, but also a common heritage and identity, similar lifestyles, and similar feelings about the world and how it ought to be. Each member of a particular culture differentially shares in the totality of distinct traits attributable to his or her culture, but nevertheless shares a large cluster of traits in common with the culture's other members, and can usually recognize who does and does not share that same culture. Local cultures also exhibit the characteristics of a system, such that changes in one component will usually necessitate and prompt changes in other components.[4]

---

is mostly irrelevant to the concerns of this book. I prefer "cultural anthropology" because for me it is simpler to read and also seems more encompassing.

[4] My brief description of what constitutes a "culture" paraphrases the definition of "cultural system" in Orbach (1977: 1), which I gratefully acknowledge. For most readers it is easy to envision small-scale fishers as members of specific cultures and subcultures, but a few students of the fisheries report that the same is true of large-scale industrialized fishers. Orbach (1977: 1), for instance, argues that the highly industrialized tuna seiners of San Diego, California, are part of an industrial culture that shows coherence, continuity, and cohesiveness extending beyond the fishers' mere ethnic affinities originating in their home ports. A similar conclusion can also be inferred from Warner (1977, 1983), describing the international trawling fleet in the North Atlantic. M. L. Miller (1983: 301) urges that a more holistic view be taken of what constitutes a marine community. Such communities, he says, should be "viewed properly as networks of focused interest groups, policy-makers, professionals, and publics."

The Characteristics of Fishing Cultures and Peoples

Granting the considerable diversity of fishing peoples, what is generally characteristic of their cultures, and particularly of the cultures of small-scale fishers, who constitute the vast majority of fishers worldwide? What general features do such cultures have in common? In the previous chapter I mentioned several cardinal features of fishing peoples: they derive their livelihood primarily from the sea; their view of the world is essentially local; and while collectively they number around 100 million people, their social, political, and economic clout is relatively minor.

To these features we can add several more. Nearly all fishers stress independence, self-reliance, freedom from regimentation, and challenge as important aspects of their occupation. A high degree of independence and self-reliance is necessary for psychologically coping with fishing activity, and it is especially important for seagoing fishers, who are beyond the support and help of their communities ashore (see, e.g., Poggie 1980a; Poggie and Gersuny 1974; Pollnac 1976, 1988; Pollnac and Ruiz-Stout 1977). Faced with a rapidly changing and potentially dangerous force, fishers must be prepared to make critical decisions with little hesitation.

Most fishers are physically hardy people who enjoy working outdoors, and most are extremely proud of their identity as fishers, which they will sometimes emphasize even when fishing activities take up only a small portion of their total working time each year. Indeed, strong feelings of pride and satisfaction have been observed in nearly all studies of fishers, irrespective of culture and region, and several good data-based studies have further corroborated these observations (e.g., Apostle, Kasdan, and Hanson 1985; Gatewood and McCay 1988; Pollnac and Poggie 1979). Thus John B. Gatewood and Bonnie J. McCay (1988: 126) conclude from their study of job satisfaction among New Jersey fishers: "Fishermen derive a considerable 'satisfaction bonus' from their work. Fishing is not merely a means to an end, but is intrinsically rewarding. . . . Fishing is not just a livelihood, it is a way of life."

Such personal satisfaction encourages fishers to be unusually tenacious in their adherence to their occupation, which sometimes puzzles economists and fisheries managers confronted with the fact that fishers will persist in the pursuit even in the face of diminishing stocks, declining yields, and very substandard incomes. As Arthur F. McEvoy (1986: 69) notes: "Fishing requires special skills as well as a tolerance for hard and dangerous work at low pay. It also has the power to hold the loyalty of its workers and their children, who will to the consternation of modern economists stay in the business long after it ceases to produce incomes comparable to those in other trades."

In communities where fishing is the main occupation, it will always be interwoven throughout the fabric of the local culture. In such communities it will also be the central attribute of the community's identity. It will pervade important rituals as well as the main social and economic institutions, and it will be the subject of popular myths, folktales, and local history. The various surrounding conventions and mores may have been handed down through several generations. Thus in communities composed mainly of fishers or in which fishers constitute a large proportion of the local populace, fishing will usually be considered as much a way of life as a way of making a living (see P. Thompson et al. 1983).

Compared with those who do not fish for a living, fishers are usually more mobile, especially geographically, and sometimes economically as well. They are often able to enter a particular fishery rather quickly when new opportunities arise. Such mobility often poses special problems for fisheries managers—problems with few analogs in the management of most land-based resources, where common-property ownership and open access are far less commonplace.

An overwhelming majority of the primary producers in the fishing industry are male, which has important implications for patterns of social and economic organization and patterns of interpersonal relations, as well as for fisheries management. Just why fishing is so overwhelmingly a male occupation bears mention here. Mainly it seems to stem from the disproportionate share of child-rearing responsibilities that women assume in practically all societies. Fishing vessels have acute space limitations, and there is rarely extra room for nonfishers, especially not babies and undisciplined young children. Moreover, it would make little sense to needlessly expose families to the considerable hazards associated with ocean fishing. Thus we find reinforced a division of labor by sex that has men working aboard fishing vessels while women serve as the mainstays of home and community life. Of course, once this pattern becomes institutionalized it may also become less flexible than it actually needs to be. In many fishing societies, adult women without children, or whose children could be cared for by others, could be very productive workers aboard fishing vessels but may be prevented from that role by longstanding and unquestioned social conventions.

Where women do engage in fish production, it is usually as shellfishers or collectors of other marine fauna along the seashore or in tidal pools—

that is, in activities that do not take them far from their children (Murdock and Provost 1973). Where they do work aboard fishing vessels, it is usually as "day trippers," that is, for comparatively short intervals of time, so that they are not away from their children for long.

There are notable exceptions to this general pattern, of course, particularly in Asia, where whole families live and work together aboard fishing boats (Ward 1955), as well as aboard certain Russian factory ships, where many women work as fish processors (Warner 1977). Moreover, while the number of female fishers remains small in absolute terms, women are increasingly finding work aboard fishing vessels in the more modernized Western nations, as a result of women's changing roles in their nations' economies. (For discussions of female fishers, see, for example, J. Van Maanen et al. 1982, on the United States; D. L. Davis 1983, on Newfoundland; and P. Thompson 1985, on Western Europe and Scandinavia.) Nevertheless, female fishers are exceptions to the rule, and this has important implications for the overall character and organization of most fishing societies and cultures, as well as for fisheries management (Pollnac 1988: 27).

One should not conclude, however, that women in fishing societies play secondary or insignificant roles in their local economies. To the contrary, as I shall shortly explain, they often make exceedingly important contributions in marketing and distribution, besides being the mainstays of community social organization and social life. Moreover, because the men are so often away and the women must take on proportionally more responsibilities in the community, it is no surprise that fishermen's wives are often more independent and are accorded relatively greater prestige in fishing communities than the women in nonfishing families. This phenomenon has been reported in such disparate regions as south India (K. Norr 1972), West Africa (H. Gladwin 1970), New England (Danowski 1980), Canada (Davis 1983), Great Britain (P. Thompson 1985), Taiwan (Diamond 1969), and Japan (Norbeck 1954). Moreover, in fishing societies where the men are absent for particularly long periods of time, there is often a pronounced tendency toward matrifocality in local social organization (Pollnac 1988: 31).

Another important characteristic of fishers is that most have a highly specialized and intimate knowledge of the marine ecosystems they exploit, although the depth and extent of this knowledge is usually much greater among small-scale fishers. When it comes to formal education, however, fishers are often at a disadvantage compared with nonfishers. Many young men have been enticed into quitting school and taking to the sea, which may be part of the reason why fishers are so often held in low esteem by their neighbors.

In fact, being held in low esteem by their nonfishing neighbors seems to be a rather ubiquitous phenomenon for fishers in many societies and cultures around the world.[5] Sometimes even their nonfishing kin and neighbors share such pejorative views. And many modern urbanites who know very little about fishing often view fishers as pariahs or as members of an underclass. Even in certain trade publications and technical journals, particularly those catering to large-scale, industrialized fishers, small-scale fishers are sometimes described as backward, and in certain fisheries management contexts they are discussed as lawless poachers or as obstacles to sound management. Thus fishers sometimes become the scapegoats of failed management policies.

Just why fishers are so often held in low esteem is a matter of some conjecture, but that they are can hardly be doubted. Their isolation from modern society is certainly part of the reason. But, it may also be a result of their adamant individualism, the clannishness of their communities, and their characteristic suspicion of strangers.

Fishers' propensity for risk taking and their relatively unsettled lifestyle undoubtedly contribute to the problem of low esteem, as do certain other characteristics—for example, their supposed disinclination for deferred gratification, such as savings and investment. Many nonfishers are critical of fishers because they make their living by "capture," rather than by long-term investment and the development of natural resources. In an impoverished Mexican coastal community in which I have spent much time, farmers indignantly describe local fishers as "robbers of the sea, who take our national resources while putting nothing back."

Some researchers point to fishers' relatively heavy use of profane language and tendency toward aggressive behavior, brawling, and physical violence as the key to others' dislike. Still other researchers argue that the widespread criticism of fishers stems from their "macho" attitude and behavior, an argument that seems to disregard the risk taking and individual daring necessitated by this particular occupation. Nevertheless, this line of thinking does have some merit, in that fishers are often socially inex-

---

[5] On the low esteem in which fishing peoples are held, Nishimura (1973: 5) concludes, from the results of a survey of 66 specialists in maritime ethnology, that "fishermen's communities, being less privileged, are often separated, isolated from, and despised by other communities even within a homogeneous society, ethnically and culturally."

perienced younger men—adolescents or men just out of adolescence—who are attracted to the occupation for its adventure and freedom from social restraint, and who are the most likely community members to engage in "macho" behavioral displays.

Certainly an important factor contributing to the low opinion of fishers is their relatively high incidence of alcoholism and disruptive influence in community life when they are home. In many fishing communities, practically the only activities that nonfishers see fishers engaged in are those associated with relaxation, leisure, play, and recreation—not work. Similarly, many community members complain about fishers' indifference to community affairs, or point to their abandonment of families or their formation of additional families elsewhere.

Moreover, the images associated with the lifestyles of many fishers often do not juxtapose favorably with the images of modernization and modern people so prevalent in the late twentieth century. Even in the developed nations, many small-scale fishers still employ rather rudimentary technologies, live in ramshackle homes with fishing gear scattered about, dress in well-worn clothes, and have few modern conveniences. One of my students once suggested that the widespread contempt for the fishing occupation arises from nothing more than the nature of the work, which is often nocturnal, always dirty, and usually smelly.

Fishers share a somewhat paradoxical worldview and ethos that separate them from nonfishers living nearby who otherwise may share their culture. This is partly because fishers are often not as well connected to the modern world as are nonfishers from the same region. Coastal communities are often geographically remote from the noncoastal communities in their regions and even from one another because of the linearity of coastlines, such that communities tend to be strung out rather than clustered around a major population center or distributed evenly throughout a region.[6] Geography also isolates coastal communities from the modern world in a psychological sense. The greater vulnerability of coastal communities to outside incursions, especially economic incursions, often prompts coastal dwellers to be more suspicious of outsiders than are noncoastal dwellers in neighboring communities.

The disarticulation of fishers from modern society is also caused by their periodic estrangement from the full web of social life in their own communities as a result of the time they spend at sea. This is particularly

[6] Hewes (1948) discusses the isolating effect that the linearity of coastlines has on fishing communities.

true in the case of large-scale, industrialized fishers. The fishing trips of the tuna seiners of San Diego, California, for instance, usually last around 40 to 60 days, and typically these fishers spend an average of 8 or 9 months a year at sea, away from their homes and families (Orbach 1977: 23). Even more estranged are distant-water fishers, such as those working aboard the factory ships in the North Atlantic, who may be continuously at sea for up to a year or more (Warner 1977). Fishers may wonder about events going on in their home communities that are affecting family members and friends, yet they are frustrated because they know there is little they can do to influence them.

Generally speaking, a fisher's life—particularly the life of a long-voyage fisher—is more limited in scope than that of a nonfisher. As Orbach (1977: 25) notes of the tuna seiners, the environment they live in while at sea is quite restricted with respect to space, alternatives for social interaction, sources of relaxation and amusement, and sensory stimulation. Moreover, in the crowded living conditions characteristic of most fishing vessels, achieving true privacy is nearly impossible.

Fishers' isolation may also stem from a personal preference for working in a more simple social environment, constituting a rejection of the complex demands and social restraints made on individuals who live in land-based communities. Moreover, as mentioned earlier, many small-scale fishers in the developing nations are victims of the marginalization that often accompanies development and modernization. Whatever the underlying causes, this isolation from the modern world greatly contributes to fishers' relative inability to influence fisheries policies, an inability that is often detrimental to their general welfare.

Paradoxically, while fishers may not be particularly well integrated into modern society, in certain other ways they are usually more worldly and less provincial than their nonfishing kin and neighbors. This is because their fishing activities sometimes take them away from home and bring them into contact with other peoples from distant communities.

These marked differences between fishers and their nonfishing neighbors are most pronounced in tribal and peasant societies. In these simpler types of societies, two distinct local subcultures sometimes coexist in the same community, one consisting of fishers and their families, the other consisting of the nonfishing populace, which is usually involved in agriculture, small businesses such as shopkeeping, and wage labor. That the two subcultures are often so distinct within a single community is particularly remarkable considering that fishers and their nonfishing peers are often kin as well as neighbors.

Dangers and Coping Responses

The sea is truly, as Walt Whitman wrote in *Leaves of Grass*, a place of "unshovelled, yet always ready graves." Fishers confront special risks and uncertainties, many of which have few parallels for nonfishers, and which over time produce a distinctive set of attitudes, behaviors, and worldviews. Like hunters of wild animals, fishers have little control over their prey, and they have even greater difficulty asserting rights of ownership or access to it because most living marine resources are common property.

As we have seen, the differences in ethos between fishers and nonfishers stem from many factors, but perhaps most importantly from the extreme psychological adaptations necessitated by wresting a living from the sea. In other occupations, to be sure, one may also be injured at work, fail to secure food or earn an income, or suffer financial reversals. However, few land-based occupations present individuals with the risk of losing all of their productive capital—as well as their lives—every time they go to work. One of the biggest dangers faced by fishers is, of course, having their vessel sink at sea, and it must be stressed that even the largest and most technologically advanced fishing vessels used today are still sometimes lost in violent ocean storms.

Although fishing is decidedly more dangerous for fishers aboard smaller vessels, especially those that are poorly equipped and maintained, fishing aboard large-scale, industrialized vessels is still quite dangerous for all crew members who are actively involved in setting fishing gear and bringing the catch aboard. Being swept overboard in extremely rough or cold water, and working in close proximity to heavy fishing gear, equipment, and other machinery, constitute the main risks (Orbach 1977; Warner 1977). William W. Warner (1983: 73) describes the hazards of working aboard the huge factory trawlers that ply the North Atlantic: "The kind of work which fishermen routinely perform in bad weather has no parallel in the world of seafaring. On a merchant ship such feats as leaping to catch wild cables or ducking out of the way of objects shifting around on deck are normally required only in emergencies, but aboard trawlers they form part of nearly every shooting or hauling of the nets."

Thus even in the world's most modern and developed countries, the fisher's occupation is still perilous indeed. British fishers, for instance, have a fatal accident rate over twenty times that of British workers in manufacturing (P. Thompson et al. 1983), and this discrepancy is essentially the same in the United States, where safety standards for fishers are generally among the highest in the world. Concerning the United States, John J.

Poggie, Jr. (1980b: 123) observes: "Official statistics affirm the extreme risk involved in fishing. Indeed, fishing is far more dangerous in terms of loss of life than coal mining—the most dangerous landbased occupation in American society. The Office of Merchant Marine Safety in 1972 reported that in 1965 the commercial fisheries of the United States recorded 21.4 deaths per million man-days in contrast to 8.3 in coal mining." In describing the perils faced by New England fishers, Poggie and Pollnac (1988: 75–76) report that "despite the advanced technology of modern vessels and survival gear, lives are lost in the most frightening of circumstances. . . . The speed with which boats sometimes capsize and sink is phenomenal. Although adequate survival gear is carried by many vessels, there is little time to put it on when the boat is listing in heavy weather." Thus one has only to imagine how dangerous marine fishing is for other, less well-equipped fishers, those in the developing nations, for instance, who cannot afford modern fishing technologies and safer lifesaving gear.

Because of the dangers inherent in working at sea, fishing usually takes on a heroic cast in fishing communities and is usually regarded as more exciting than most other local occupations. Thus it is often seen as more desirable, even when it yields lower incomes than nonfishing occupations. As noted earlier, while in many rural communities fishers may also work at agriculture, animal husbandry, or wage labor jobs, they often stress the primacy of their occupational identity as fishers, even when for most of the year they are engaged in nonfishing activities. Particularly in simpler and more rural societies, fishers often manifest many of the psychosocial attributes of gamblers: a great predilection for taking economic and personal risks, an emphasis on individualism and personal autonomy, a desire to be socially unconventional, and a need for excitement.[7]

The renowned navigators from Puluwat atoll in Micronesia, for example, sail homemade canoes across vast ocean distances while using neither maps nor other forms of modern navigational gear. Although they

[7] Interesting observations comparing the bravery, adventurousness, interpersonal aggression, and other personal attributes of fishers and their nonfishing peers in the same communities are found in a diverse literature, including the following: on aggression, Aronoff (1967); on "rough language," Glacken (1955); on fishers' drinking, T. Gladwin (1970); on "macho" attitudes and behaviors, Andersen and Wadel, eds. (1972), Pollnac (1976: 67–68), and Tiller (1958), among others; and on the heroic quality of ocean fishing and the propensity for taking risks and for individualistic daring, e.g., Abrahams (1974), Bernard (1967, 1972), Forman (1970), Fraser (1960), T. Gladwin (1970), Poggie and Gersuny (1974), and Pollnac and Ruiz-Stout (1977). Whether fishers manifest a lower propensity for deferred gratification is discussed in Poggie (1978); Poggie, Pollnac, and Gersuny (1976); Pollnac and Poggie (1978); and Poggie, Bartee, and Pollnac (1976).

have an abundant fishery immediately around their island community, they often make long ocean voyages to a distant, uninhabited island to harvest sea turtles, which they consider a delicacy. According to their principal ethnographer, Thomas Gladwin (1970), these voyages are undertaken not so much to secure food as to reaffirm values central to their cultural identities as heroic men of the sea. No doubt their voyages would be perceived as unthinkably dangerous by their nonseagoing neighbors, as well as by most modern urbanites.[8]

The particular risks and uncertainties associated with marine fishing have prompted—among fishers of all types—the development of a rich variety of superstitions, magico-religious mythologies, taboos, rituals, and other beliefs in things magical and supernatural (see Anson 1965; Dorson 1964; Creighton 1950; Frazer 1890; T. Gladwin 1970; Goode 1887; Orbach 1977).

Even among modern, industrialized fishers, the observance of various taboos and acting out of certain magical beliefs are quite commonplace. Warner (1983: 78), describing fishers aboard West German factory trawlers in the North Atlantic, states that "the West Germans are almost as superstitious as other fishermen. One of their principal concerns is Sunday, a day that is widely held to bring ill fortune." Similarly, Orbach (1977: 210–11) notes that the tuna seiners he studied—who come from as modern a setting as San Diego, California—profess to believe that growing beards will lead to longer trips, crossing eating utensils in the galley may bring bad luck, whistling on the bridge can bring wind and bad weather, and abusing or not paying a prostitute may bring about equipment failure. Orbach suggests that such beliefs are bandied about partly as a means of coping with boredom, but more important, he joins many others who feel that the main reason such beliefs develop is that they help fishers cope with the various anxieties arising from fishing activity (see Homans 1941; Jahoda 1969; Kluckhohn 1942; Malinowski 1954; Poggie and Pollnac 1988).

In his studies of the Trobriand Islands, the renowned anthropologist Bronislaw Malinowski (1954: 31) noted that practically no superstitious beliefs or ritual behaviors were associated with comparatively low-risk lagoon fishing activities, whereas quite the opposite was true with deep-sea fishing activities. Likewise, E. G. Burrows and M. E. Spiro (1953), who

---

[8] Puluwatan navigation, watercraft construction, sailing techniques, and fishing methods are thoroughly described in T. Gladwin (1970).

studied local societies on the Pacific island of Ifalik, observed no magico-religious rituals performed in association with farming activities, but noted a wealth of such rituals in association with ocean voyages and canoe construction. Similarly, William A. Lessa (1966), in a study of the maritime people living on the Micronesian island of Ulithi, found no rituals performed in conjunction with inshore shellfishing and short ocean voyages, but extensive ones performed in conjunction with ocean fishing and long ocean voyages.

Similar observations have been made of modern fishers in the Western world. John J. Poggie, Jr., R. B. Pollnac, and C. Gersuny (1976), for example, in a study of modern New England fishers in the United States, observed lower degrees of magico-religious beliefs and rituals among day-trippers than among fishers whose trips usually took longer than one day.

The foregoing studies describe magico-religious beliefs and behaviors that can be interpreted as coping responses arising from concerns for personal safety, but fishers clearly also develop such beliefs and behaviors to help them cope with fishing's economic uncertainties (see Prins 1965; Watanabe 1972; Oto 1963; Price 1964; Orbach 1977). This has led a few researchers to wonder whether economic risks are not ultimately more influential in the development of fishers' magico-religious beliefs and practices than the threats to personal safety (see Mullen 1969; Lummis 1983, 1985). In essence, they argue that threats to personal safety are more or less uniform among fishers, whereas threats associated with possible economic outcomes are more variable and less predictable.

However, Poggie and Pollnac (1988: 73), in a controlled study analyzing data gathered in three southern New England ports, conclude that "the principal function of ritual avoidances among fishermen in Southern New England is to reduce anxiety resulting from uncertainty with respect to personal safety." These findings reaffirm a conclusion Poggie had come to several years earlier: ritual observances and taboos are responses to "the perceived risk associated with protection of life and limb, and not with production of fish, a distinction not made by many theorists" (1980b: 124).

As I shall discuss in Chapter 8, local folk beliefs about "luck," which explain differential success among fishers who are otherwise peers, are often developed out of a need to maintain local myths of equality. In this instance, such beliefs can be seen to arise from the need to cope with the risks and uncertainties posed by social life.

Patterns of Work and Social Relations

The special risks and uncertainties posed by fishing activity also prompt significant differences between fishers and nonfishers in the nature of income, remuneration schemes, and other economic arrangements. Such differences have important ramifications for patterns of interpersonal relations and for local social and economic organization. For example, incomes deriving from fishing activities, even when comparable to or overall better than those in local agriculture, are almost always less predictable and more sporadic. In a typical season a fisher may experience everything from prolonged poor catches to sudden windfalls of cash proceeding from extraordinary catches. This fluctuation prompts a higher dependence on debt peonage among fishers than among nonfishers. Local storekeepers extend credit by advancing fishers food for their families, while local intermediaries advance cash for family support, gear, fuel, and so forth, at the same time locking up exclusive rights to market fish catches later on.

Remuneration schemes also differ markedly from those for nonfishers. Fishers rarely work for hourly wages. A few are compensated by means of "count systems," in which they receive payment on the basis of the total number of fish landed, but most are compensated on the basis of predetermined shares from the proceeds of catch. This "share system," which is also sometimes referred to as the "lay system," is the most ubiquitous means of compensating fishers around the world and is equally common in simple and modern societies.[9]

The share system promotes a greater emphasis on cooperative behavior among fishers, particularly among those who work together aboard the same fishing vessel, than is required in most work situations. Moreover, it usually fosters a spirit of egalitarianism among fishing work groups, since, as Pollnac (1988: 30) states, the share system "enhances each individual's perception of himself as being a participant in a common endeavor." Orbach (1977: 182) similarly comments that fishers who participate in a share system are "co-adventurers," whose livelihood depends on a common endeavor. On the negative side, however, the share system makes income production among fishers far less certain than among agricultural

---

[9] Andersen (1988: 99) explains how the count system of compensating Newfoundland's bank-schooner-and-dory fishers during the late nineteenth century and the early twentieth century "drove men to take heavy risks with their lives, as in overloading their dories." A whaler's perspective of the share or lay system appears in ch. 16 of Herman Melville's *Moby Dick* (1851).

workers, who are usually guaranteed a return based on the number of hours they work or on the quantities of crop they harvest.

Because of the need for highly coordinated teamwork at sea and the ever-present problem of low incomes, kinship is often a strong factor underlying the recruitment and composition of fishing vessel crews. This is especially true among small-scale fishers, but even in large-scale enterprises vessel crews often consist of at least loosely related kin groups. There is a great need to maintain harmony and cooperation among crew members at sea, something usually better established within kin groups than with random groups of strangers. Furthermore, by limiting crews to kin groups there is a better chance of keeping income within these groups.

There are exceptions, of course. Fishers in Okinawa, for example, work on different vessels in order to minimize the potential loss to families should a vessel be lost (Glacken 1955), and shark fishers along Mexico's Pacific Coast avoid hiring relatives, who might resent taking orders from family members (McGoodwin 1976). Nevertheless, as Pollnac (1988: 29) notes, in most world regions kin ties are used to recruit fishing crews, both to ensure harmony and to keep the proceeds of the catch within certain kindreds.

Because most primary producers are male, and because their fishing activities often require them to be away from their families and communities, their prevailing patterns of kinship relations, friendships, and other community associations also differ markedly from those of their nonfishing counterparts. There is a curiously unique and dichotomous character to their patterns of social relations. As we have seen, this dichotomy entails, on the one hand, intense relationships between men at sea, strong male bonds, cooperativeness and camaraderie achieved during prolonged periods of virtual isolation in close quarters while sharing physical and economic risks, and on the other hand, tenuousness, conflict, estrangement, and ephemerality in important interpersonal relationships ashore.

Many peoples in the developing nations, particularly agrarian peasants, have been described as adamantly individualistic and reluctant to join in cooperative endeavors. These normatively uncooperative attitudes and corresponding behaviors have often been seen as major factors inhibiting rural development. Boat crews among peasant fishers stand in marked contrast. As mentioned above, not only is the share or lay system conducive to cooperation, but the dangers inherent in work at sea—where everyone is literally in the same boat—considerably discourage the sort of uncooperative attitudes so often reported among peasant fishers' nonseagoing neighbors.[10] In the words of Joseph Conrad (*Heart of Darkness*), "Between us there was . . . the bond of the sea. Besides holding our hearts together through long periods of separation, it had the effect of making us tolerant of each other's yarns . . . and even convictions."

Unlike their counterparts in the large-scale sector, most small-scale fishers are day trippers. Nevertheless, many are often absent from their communities for long periods of time when they migrate to seasonal fishing camps or to other fishing communities in order to take advantage of opportunities distant from their home ports. While residing in other communities, they occasionally form important new social ties, sometimes even starting up additional families. These new ties compete with those in their original communities, contributing still further to the conflict and tenuousness that so often characterize fishers' primary social relationships.

When fishers return to their home communities after prolonged absences, their presence is often disrupting to the established, ongoing social order. Returning home, they greatly desire rest, relaxation, and immediate intimacy, but they are out of sync with the rhythms of daily life in their communities. They recount their sea stories as a means of underscoring for their family members that they indeed had an existence while they were away, but soon the family members tire of hearing them. They may indulge their children, breaking down established patterns of household discipline, or make extravagant purchases for themselves or other family members while little appreciating the household's overall financial situation. As the primary producers of household income, they may attempt to assert their dominance as the head of the household, oblivious to the established patterns of daily life in which other, permanently resident adults have been the day-to-day managers. Thus as household discipline breaks down, as the children's behavior becomes more problematic, and as arguments erupt with spouses and other household members, fishers often feel alienated, betrayed, and perplexed by the difference between the images they had of their homecoming during their long time away and its reality.

Looking beyond their households for sympathy and support, fishers often find little in common with other, nonfishing members of the community. To them, fishers are practically unknown, nonentities. Fishers may

---

[10] McGoodwin (1979) discusses differences in behavior among small-scale fishers from the same community in Pacific Mexico: individualistic, secretive, and uncooperative while fishing inshore; cooperative and friendly while fishing offshore. The article stresses the importance of context, rather than local cultural norms, in producing behavioral and attitudinal differences among people who are otherwise from the same community and culture.

also find life ashore more complicated and fast-paced than the life they knew at sea, which makes them feel unsure of themselves in social situations. This in turn may lead to a timidity in social relations that is in marked contrast to a more exuberant mode of relating to fellow fishers while at sea. Thus they are inevitably drawn back to the company of their fishing comrades, who find themselves in similar situations. As Orbach (1977: 272) notes, "The one audience to whom a fisherman can play without restraint is another fisherman." So the fisher seeks out other fishers, perhaps goes on drinking binges with them, and unrealistically idealizes these social ties, which had been found so limited while at sea. Soon the fisher begins to plan or anticipate the next fishing expedition—now motivated as much by escapist desires as by economic necessity.

Over the long term the difficulty fishers have in integrating themselves into social life ashore, and particularly into family life, often prompts the development of chronic psychological problems. These are manifested in a variety of ways, including antisocial behavior, ambivalent feelings regarding close social ties, distrust of significant others, feelings of guilt at being away so much, alcoholism, drug abuse, and so forth. Even being at home, which should be a time for reunion, as well as for rest and relaxation, can induce ambivalent feelings, since fishers often feel economically impotent and useless when they are not earning an income. Moreover, these problems are aggravated by the prolonged periods of isolation and social and sensory deprivation fishers experience after they return to sea.

A seafarer's social relations, characterized by loneliness and estrangement while at home and only limited possibilities while at sea, are lyricized by an anonymous English poet in "The Greenwich Pensioner":

> Twas in the good ship Rover
> I sailed the world all round
> and for three years and over
> I ne'r touched British ground
> At length in England landed
> I left the roaring main
> Found all relations stranded
> and went to sea again

Fishermen's Wives

For fishermen's wives, the fisherman's pattern of social life produces similar strains. While husbands are away at sea, their wives are equally plagued with loneliness. They worry about their husbands' welfare, long for their return, and cope with the uncertainty of their households' future income. Moreover, in much the same manner as the men, they often develop especially close relationships with those around them, in this case with other women, and they also often become the principals in community affairs. Thus when the fishermen return, the women may find their presence in the community disruptive. Soon the women may also find themselves wishing for the men's departure on another fishing expedition. "Can't live with 'em, can't live without 'em," certainly could have been coined by a fisherman's wife.

An anonymous letter from a fisherman's wife, published in *National Fisherman* under the title "Warning: How to Lose a Family" (1988), evokes a painful portrait of the dislocations of family life. First the author implores the magazine's editor to consider the following:

> These tough guys are real people with real families, and more than a few have real serious drinking, drug and emotional problems. . . .
>
> Whether fishing produces these flaws or draws them to it, I don't know. It's not the time and distance that destroys these guys' families; it's the hard-core attitude when they hit land. They're very loving on the sideband.[11]

Then, addressing her comments to fishermen themselves, she continues:

> We don't think you know how much we appreciate the hard, exhausting and dangerous work you do, or that we understand what your blood, sweat and tears produces. We're thankful for our home, our treasures and the food we eat.
>
> We don't know how to reach you anymore. Our lives are as vast as the sea from which you fish. We want to love you and respect you. . . .
>
> Every time you leave, you take other people's souls with you. We sit looking at the full moon, wondering how and where you are. The kids draw pictures of your boat and tell everyone with such pride in their eyes, "My daddy's a fisherman."
>
> We spend our time taking care of your home, so it will be clean and warm when our ship comes in. We realize it's not an easy transition from your world to ours, but if you will trust us and let us, we will help you make the adjustment.
>
> After all, we're hooked on you, and we are the best catch you ever hauled in.
> How to lose a family:
> —After a 21-day fishing trip, spend the first night home with the guys celebrating your catch.
> —When you finally do make it home, don't forget to accuse your spouse of messing around with every man left on land while you were gone. You could even slap her around, just to be sure she won't do it next time you're gone.

[11] A sideband is a marine radio transmitter/receiver.

—Criticize the kids for the way they mowed the yard and painted the garage. Why should you thank them? You pay the bills.

—Get good and drunk the night before your only day off. That way you'll sleep soundly the next day, and you won't have to worry about being amorous.

—When you finally do wake up, don't play with the kids; they will be much more fun when they're older. Besides, you haven't watched a game on TV with your buddies lately.

—Send your wife to the store on a beer run, a cigarette run and a pizza run. After all, what's she there for anyway?

—Spend all your money in the bar before you leave. Then you won't have to give any to the rug rats for summer camp.

While a fisherman's wife is certainly the mainstay of his home and family life, and often of his community's social life as well, the literature on fishing communities is also replete with testimonials to her importance in local economic affairs. As has often been observed, these women are frequently the principals in the first phases of distribution of their husbands' catches. Typically, they take part of the catch for their immediate household needs and some for bartering with relatives and neighbors, and then are responsible for marketing the remainder. This is congruent with observations in many fishing societies that fishermen are generally uninterested in distribution and marketing activities, considering their work to be at sea. For many fishermen, the home port is a place where one returns for reunification with family, kin, and friends, and seeks relaxation and recreation, not more work.

The life situations of fishermen's wives and their central role in the economic and social affairs of their communities contribute importantly to the distinctive character of fishing cultures. Within fishing communities, the roles of women from fishing families often differ markedly from those of women not from fishing families, and even more markedly contrast with the roles of women in nonfishing communities nearby. Moreover, as noted earlier, many fishing communities manifest tendencies toward matriarchy and matrifocality, that is, an organizational pattern in which women are preeminent in the important affairs in community life, and in which normative household compositions consist of a woman and her children, which is often extended to include the woman's mother, her sisters, and her sisters' children.

The central role women often play in distributing the catch has been commented on in many ethnographic accounts of fishing societies, but few studies have gone on to consider the broader implications of women's importance in all the other aspects of community life. Indeed, discussion of the vital role of women in fishing communities is probably the greatest lacuna in the available literature on fishers—one that needs to be filled if we are to have a more comprehensive view of the cultures of fishing peoples.[12] Moreover, learning more about women's roles in fishing communities will be essential for developing a more humanized approach to fisheries management.

Fishers' Knowledge of Marine Ecosystems

Many students of fishing have commented on how thoroughly fishers understand the marine species and ecosystems they exploit. Even modern, industrialized fishers, of whom it has often been said that "most of their skill is in their machines," usually manifest a significant understanding of the marine environments they exploit (see, e.g., Orbach 1977: 72–103; Warner 1983: 167). Unquestionably the seafaring life offers plenty of time for observation and reflection. Thus Warner (1983: 167) says of fishers working aboard the giant factory trawlers in the North Atlantic, "As a result of their constant observations, quite naturally, most distant water fishermen become excellent empirical biologists."

However, far more remarkable in terms of the depth and extent of practical understanding, as well as the relative degree of intimacy with particular marine environments, is the marine biological knowledge of small-scale fishers. John Cordell (1974) reveals, for example, how impoverished swamp-dwelling peoples in Brazil's Bahia exploit an extremely complicated estuarine ecosystem, employing an esoteric knowledge of its geographic and hydrographic attributes, some of which are very subtle. Their knowledge of the tidal cycles and how these affect the productivity of various submerged microenvironments is used to devise capture strategies and select fishing gear for the various times of the lunar day, month, and year. This specialized knowledge, Cordell argues, which stems from these fishers' long-term association with the ecosystem, is the main factor permitting them to subsist in their fishery even though it has been invaded by technologically more sophisticated commercial fishers, who thus far do not have this specialized knowledge.

The marginalized rural fishers I studied along Mexico's Pacific Coast, who work in an environment very similar to that described by Cordell in Brazil, have developed complex patterns of exploitation that are remark-

[12] This subject is slowly being addressed, and many new studies focusing on women in fishing communities are now appearing. See, for example, Maréchal, ed. (1988); and Allison et al. 1990.

ably similar. They have developed other techniques as well that attest to their intimate understanding of the marine ecosystem. They taste mouthfuls of brackish water while fishing in the estuaries, for example, for clues about which fish species are present, and they also submerge their heads to listen to the clicking of the shrimps in order to determine whether they are holding fast in the sea grasses along the estuarine floor or moving in the water column.

Similarly, the Cha-Cha fishing peoples from St. Thomas, Virgin Islands, have developed an "ichthyology" all their own, with a utilitarian purpose. Their classificatory system distinguishes fish not so much on the basis of morphological characteristics as on the basis of their behavior, particularly behavior relating to their capture. This system also assesses the likelihood that certain species and the marine environments they are caught in will carry *ciguatera*, a deadly poison (Morrill 1967).

Folk peoples' intimate association with natural environments is often seen as exotic by modern urbanites, who lament their own estrangement from nature. It is also seen as very important among maritime anthropologists concerned with recording traditional culture-and-environment relationships before they are lost forever through acculturation. Highly specialized and intimate knowledge of marine ecosystems cannot be presumed to exist among all small-scale fishers, however; it is largely confined to those with a long tradition of fishing in a particular locale, a tradition going back at least several generations.

What is more generally true of small-scale fishers is that they are more concerned with whole or entire marine ecosystems, and with a wider and more diverse range of fish species, than are most large-scale, industrialized fishers. As a result, these two different types of fishers will encourage different emphases in fisheries management policies. And because the components of a marine ecological system are interrelated, fisheries policies that are primarily responsive to the broader needs of small-scale fishers will generally go further toward maintaining the overall health of a marine ecosystem than those formulated primarily in response to the narrower needs of large-scale fishers.

These different requirements of the two groups derive in part from their characteristically different means of distributing their catches. Industrial-scale fishers are basically oriented to production for the market, including the world commodity market, whereas many small-scale fishers produce only for household subsistence, community food needs, and other local markets. The small-scale fishers' greater preoccupation with subsistence makes them inherently more interested in ensuring the availability of a variety of seafoods, since variety makes for a more nutritious and interesting diet. These concerns ultimately serve conservationist principles, since the maintenance of species diversity is usually seen as a fundamental requirement for the maintenance of a healthy marine ecosystem.

Small-scale fishers, particularly those described as artisanal, have been referred to as "Ecosystem People" (Dasmann 1974). As Gary A. Klee (1980: 1) explains, this descriptive label implies members of indigenous or traditional cultures "who live within a single ecosystem, or at most two or three adjacent and closely related ecosystems." "Biosphere People," by contrast, are tied in with global markets and the employment of more sophisticated and effective technologies. The first tend to be aware that if they deplete their main subsistence resources they risk their own ruin; the second, to subscribe to a "myth of superabundance," the feeling that there are always other ecosystems and other resources to exploit should the ones they currently favor run short.

For fisheries management the difference between ecosystem and biosphere fishers is crucial. Ecosystem fishers, because they are relatively less mobile and are dependent on only one or a limited number of marine ecosystems, are more vulnerable to the unfavorable consequences that would ensue from the depletion of those few systems. Biosphere fishers, after depleting a particular marine ecosystem of its valuable resources, may merely redirect their efforts elsewhere. This nomadic, global approach to fishing, which gained considerable momentum following the close of the Second World War, had become so widespread by the late 1960's that it was a prime factor in the leveling off of the world's fish catch, which was first noticed soon thereafter (Maiolo and Orbach, eds. 1982: 63). By then huge fleets of factory trawlers—some consisting of more than a hundred vessels—roved the seas, wiping out whole stocks in some ocean regions and then moving on to new grounds (Pontecorvo 1986: 7).

The literature on fishing peoples stresses that traditional peoples who are dependent on local ecosystems get to know those systems intimately, and because such peoples are less aware of the outside world and its multiplicity of other ecosystems, they do not behave as if important food resources are available in unlimited supply. Instead, over long periods of time their intimate association with the environment has been conducive to their inventing conservation measures that could teach much to the modern manager. As Klee (1980: 1) states: "Primitive or traditional cultures have insights regarding living with the earth that the technocratic

world has lost. Modern resource managers should drop their superior attitude and take a closer look at what these societies did to conserve resources."

While statements like this are commonly encountered in studies of fishing peoples, they are often asserted too broadly and are seldom supported
by quantitative data. Moreover, they sometimes seem to reflect an unrealistically romantic, "Rousseauesque" view of fishing peoples.[13] Thus we
should regard such statements with a degree of caution. As a fisheries management report of the FAO (1983: 11) reminds us, "Locals have had more
incentive to self-regulate a particular fishery than have nomadic roving
fleets. However . . . even locals can over-exploit a stock if there is not adequate social control of the number of local participants."

It is important not to extend claims such as Klee's, which pertain mainly
to primitive or traditional peoples for whom such claims *are* more or less
true, to small-scale fishers in general. There may be marked differences in
the conservation practices of traditional fishers who have been residentially
stable for a long time and newcomers to a fishery. Furthermore, as we shall
see, some fishing peoples may never overfish important marine resources
because they are simply unable to, not because of any conservationist
wisdom. Nevertheless, before attempting to regulate fishing activity, it
would behoove fisheries managers to study how local small-scale fishers
understand the marine ecosystems they exploit. In this regard, regulatory
means employed by local peoples will be the subject of Chapters 7 and 8.

Small-Scale vs. Large-Scale Fishing

As we have seen, large-scale fishers are generally away from their communities for longer periods of time than small-scale fishers; as a result, they
suffer dislocating social and psychological consequences to a greater degree.
Moreover, while large-scale fishers often manifest considerable knowledge
of the marine ecosystems they exploit as well as the species within them,
their degree of intimacy is generally less than that of most small-scale
fishers.

Because of relatively greater mobility and more intense harvesting
methods, large-scale, industrialized fishing, especially when it involves

---

[13] Farmer (1981: 238), in his review of Klee, ed. (1980), states that the book should be
"prescribed reading for 'experts' who descend on developing countries charged with the arrogant conviction that the land-use systems there are 'primitive' and in need of replacement."
However, he also cautions that some of the authors in Klee's book "lean a little too far the
other way in almost Rousseauesque glorification of indigenous practices." I concur with both
of these opinions.

roving fleets of factory ships, can more quickly deplete a marine resource than small-scale fishing can. On the other hand, it is usually easier to restrain large-scale fishers in order to permit stocks to recover than it is to curb fishing effort among small-scale operators.

With new censuses and the amassing of more comprehensive economic statistics, it has become apparent that the significance of small-scale fishing has been underestimated—in terms of the aggregate number of people it employs, its overall productivity, and its relative efficiency. Moreover, as the stresses and strains stemming from industrialization are increasingly felt, some fisheries experts are suggesting that small-scale fishing will play a relatively greater role in the future of the world's fisheries.

As mentioned previously, in terms of sheer human numbers the importance of small-scale fishing looms large indeed—involving, as we have seen, perhaps 100 million persons worldwide, compared with fewer than half a million in the large-scale sector. In gross economic terms, and particularly in terms of overall economic efficiency, the contributions of small-scale fishers are surprisingly large. About 45 percent of the total world fish catch that is designated for human consumption—about 20 million tons annually—is caught by small-scale fishers. Moreover, nearly all that fish is designated for human consumption, whereas about a third of the catch of large-scale fishers goes into the production of fish meal, which is used mainly for animal feed. The catches of small-scale fishers are also more often designated for local and regional consumption than those of large-scale fishers, who are considerably more involved in production for export.[14]

Small-scale fishing is more efficient from several perspectives. On average, its capital cost per job is roughly 100 times lower than the large-scale sector's. Thus small-scale fishing is less often implicated in the problem of overcapitalization, one of the most serious problems facing many fisheries today. Furthermore, small-scale fishing consumes only about 11 percent of the total fuel oil used in all commercial fishing, but produces nearly five times as much fish per unit of fuel oil consumed as the large-scale, industrialized sector does (Thomson 1980: 3).

As astounding as the foregoing observations may at first seem, small-scale fishing makes a greater overall direct contribution toward providing food for humans than large-scale fishing does. As Conner Bailey (1988b: 108–9) stresses, this is particularly important for food-deficient develop-

---

[14] McGoodwin (1987) fleshes out a mosaic of problems associated with an overemphasis on export production in the fisheries policies of Pacific Mexico.

ing countries, since large-scale modes of fishing generate very little employment for a given amount of capital investment. In these nations, he notes, "capital is in scarce supply relative to labor as a factor of production." Moreover, the greater efficiency of small- versus large-scale fishing should be of particular concern to these mostly oil-importing nations. In fact, where large-scale fishing is appraised as more efficient than small-scale fishing, the apparent efficiency is often "exaggerated by direct or hidden subsidies to large-scale fishermen frequently provided by national governments and international development agencies."

What all this suggests is that in terms of brute economics and demographics, and their implications for general human welfare—as well as for wise resource conservation and management—small-scale fishers should be at the forefront of concern in the formulation of future policies for the world's fisheries.

Small-scale fishing differs from large-scale fishing in a number of other important ways. For instance, small-scale fishers' worldview and connection with the world are typically much more localized than among most large-scale fishers. Small-scale fishers' lives are usually centered around a community or a string of communities in a coastal region. This is quite different from the more global orientation of most large-scale fishers, who range more widely and are part of an international community of fishers at sea, and who at times even become pawns in international conflicts.

In small-scale fishing communities, though fishing may be acknowledged as an important type of business endeavor, its character as a way of life tends to make it much more than that. For the people involved in large-scale fishing, it is unmistakably a business enterprise. Thus when business goes bad, large-scale fishers usually leave the fishery rather quickly, whereas small-scale fishers will often persist in fishing, clinging to their accustomed way of life, regardless.

Large-scale, industrial fishers do have unique cultures distinct not only from those of the land-based, but also from those of small-scale fishers. However, what distinguishes them most from small-scale fishers is the nature of their involvement with their work. Though they are typically more estranged from their families and communities, and in this regard probably suffer greater problems than small-scale fishers, it is their socioeconomic organization while at sea that is most different from the typical small-scale patterns. Their relationship to their employer is usually like that of an industrial wage laborer, with labor's cost, rather than kinship or community obligations, a prime consideration underlying their employment. Kinship ties may still help a fisher get a job in large-scale, industrialized fishing, but not to the degree seen in the small-scale sector. Moreover, in small-scale fishing it is common for investors, management, and workers to know one another on a personal basis and to work closely together. In contrast, as Bailey (1987: 175) notes, in large-scale fishing "the roles of investor, manager, and worker are clearly differentiated. The result is a separation between owners and crewmen along lines of economic class interests." This produces a work environment in which management honors few social obligations to its workers.

Although large-scale fishers usually have safer working conditions and sometimes more comfortable living quarters than most small-scale fishers, overall their lives seem considerably harder, owing mainly to the more continuous nature of their work and the prolonged amounts of time they spend at sea. Quite unlike factory laborers who work ashore, fishers aboard factory ships cannot leave the factory once their work shifts have been completed. "Not since the great age of whaling," says Warner (1983: 176), "have men stayed at sea for such protracted periods . . . and no whaler ever suffered the continuous and exhausting work schedules of the modern distant water fisherman."

Large-scale fishers also more often find themselves working among crew members who are comparative strangers. Sometimes a diversity of cultural or ethnic groups may be found aboard the same vessel. Moreover, even when crew members are all from the same culture or ethnic group, they still usually come from several different communities.

To a much greater degree than in small-scale fishing, large-scale fishing enterprises are constrained by economic "bottom line" considerations. They are primarily concerned with the maximization of profit and are far more inclined to slough off unprofitable operations and employees should fishing activities become unprofitable. Their loyalty to particular fisheries, as well as to the neighboring communities, is therefore relatively tenuous, with the result that few large-scale fishers feel any personal identity with a particular fishery. This is a predictable consequence of their far-ranging long-voyage operations.

The technologies employed by large-scale fishers give them considerably greater control over the marine environment. They have the ability to fish in practically any waters and in the most extreme weather conditions, as well as the ability to spend a greater proportion of their time engaged in productive activity. Most of the skill they bring to fishing is in the material technologies they employ, rather than in themselves. When it comes to locating fish, electronic gear and aircraft now often replace environmental knowledge and experience. Navigation relies on satellites in outer space, and modern communications gear provides warnings of bad weather, doing away with the need to be able to read subtle signs in the water and the sky. Further, whereas much of the fishing gear used by small-scale fishers functions passively (e.g., gill nets, fish traps, set lines), most of the gear used by large-scale fishers does not (e.g., trawls, dredges, purse seines), a fact that can and often does cause conflicts within some fisheries (Bailey 1987: 173).

In sum, large-scale fishing is mainly a business enterprise, whereas small-scale fishing is a business enterprise *and* a social and cultural enterprise—a way of life. That is why small-scale fishers will often continue to work in a fishery even when it no longer provides them any economic return. It is also what makes their management so confounding.

PART II

Fisheries Management

Start with what the people know.

Credo of the Chinese Rural
Reconstruction Movement, 1920's

THREE

Unregulated Fisheries

The first fishing peoples who ever lived did not regulate their fisheries. They did not have to. Their populations were so small and their corresponding impact on marine resources so minuscule that there was no need to constrain their fishing effort.

The First Fishing Peoples

So far as we know, the distinction of being the world's first "maritime people" (J. G. D. Clark 1948, 1952) falls to the Maglemosians, who first appeared during the Mesolithic era around 10,000 years ago. They were also the first human beings we know of who lived a semisedentary way of life, with many living in rather large settlements, compared with those of their more nomadic and terrestrial hunting-and-gathering contemporaries. Nor did they have to wander around as much as the hunter-gatherers, forced to follow wandering herds of prey animals. They mostly lived along the shorelines and subsisted by gathering wild plants and harvesting seafoods.[1]

Maglemosian settlements were not only considerably larger but also more nucleated than those of inland groups, suggesting they were able to produce food surpluses, particularly of shellfish. Their high reliance on shellfish is evident in the large shell middens they left behind, many of which stand as our earliest record of their existence. They reached their fullest cultural development around the rim of the Baltic Sea some 8,000

---

[1] Their very name, Maglemosian, conveys something of this history. It derives from *Magle mose*, or "big bog" in Danish, and refers to the group's main occupation sites scattered in the coastal marshes and bogs along the margins of the cold Baltic Sea.

years ago, near the end of the Mesolithic era, and continued to practice their way of life until approximately 4,500 years ago, well into the Neolithic era.

The Maglemosians' high reliance on seafoods is evident not only in the remains of the marine organisms they harvested—the huge shell mounds and smaller deposits of fish bones they left behind—but also in many cultural artifacts, especially various tools fashioned from bone, including barbed fishhooks and barbed or serrated harpoon points. The material remains of the Maglemosian culture also include the first "boardwalks" employed in a coastal zone. Apparently they placed flat boards around the shorelines of the marshes for ease of movement. Moreover, though not the oldest known, they also fashioned crude watercraft.

As the Maglemosians flourished, changing environmental conditions in other parts of the world encouraged the development of other maritime societies in both the Old World and the New. In the Old World, such societies were well established in Africa as early as 8,000 years ago, especially around the mouth of the Nile, and by 7,000 years ago there were fishing peoples living around the shores of Lake Rudolph (J. D. Clark 1970). These early fishing peoples left behind bone fishhooks and harpoons, shell middens, fish bones, and rock paintings depicting marine-capture activities. In what is now Japan, the Jomon fishermen-voyagers thrived around 5,000 years ago (Nishimura 1973). Maritime societies appeared in the New World around the same time as they did in Africa; coastal fishers and shellfish harvesters began to populate what is now Baja California, Mexico, around 8,000 years ago (Hubbs and Roden 1964: 145). And by approximately 5,000 years ago, fishing peoples inhabited the Peruvian coast (Moseley 1975).

Practically all of these early maritime societies exploited two adjacent and very different ecosystems—one terrestrial, the other marine—hunting and gathering in both. This afforded them greater security than their exclusively terrestrial counterparts, making them less vulnerable to short-term environmental changes and allowing them to accumulate food surpluses. Those surpluses in turn prompted ever higher levels of human population and more sedentary modes of settlement. Hence it is notable, but perhaps not widely appreciated, that the first human beings to turn from nomadism to more settled ways were maritime peoples such as these, and not terrestrial peoples.

The Dawn of Fisheries Management

Though the Maglemosians and their successors who lived throughout the Mesolithic era and into the middle of the Neolithic concentrated their food-gathering efforts in marine ecozones, there is no evidence—not even conjectural—that they practiced even rudimentary forms of fisheries management. Given their presumably low populations, their low-yield capture technologies, and the abundance of marine resources around them, it is doubtful that they had any need to.

But as ever larger and denser human populations arose in coastal areas, a point was eventually reached at which the depletion of marine resources became problematic. Fished-out areas undoubtedly prompted the first conscious perceptions by human beings that fishing effort had to be controlled, providing the impetus for fisheries management.

It may even be that the need to develop means of fisheries management was an important factor stimulating the development of civilizations in certain parts of the world. Or so one might conjecture, since the institution and implementation of fisheries management clearly require more complex modes of social, political, and economic organization than does unregulated fishing.

The first fisheries management regimes must have arisen in places where the most important marine resources were the sedentary ones, such as clams and oysters. Being essentially immobile and abundant in shallow waters near the shore, these had the least chance of escaping early fishers and thus were the most vulnerable to overfishing.

One of the earliest documented instances of severe overfishing occurred nearly 3,000 years ago along the Peruvian coast. The early coastal Peruvians achieved the status of near civilizations between 3000 and 1000 B.C., while relying almost exclusively on marine resources, particularly shellfish, and apparently without the benefit of agriculture or gathered plants. By 1000 B.C. these coastal societies were in full florescence and manifested many of the attributes scholars associate with state-level civilizations: they had several large communities, consisting of up to several thousand households, and because they were able to produce large food surpluses, many of their inhabitants were freed from having to be continually occupied with getting food, allowing the growth of specialization in craft production, trade, religion, and politics. Some of these communities constructed monumental architectural works, such as ceremonial mounds, temple sites, roads, and canals, which must have required the recruitment of large numbers of laborers. It is also reasonable to assume that these peoples lived in highly stratified societies in which a small group of elites or an aristocratic class held disproportionate amounts of power.

The Peruvian coast where these maritime societies flourished is one of the driest in the world, getting almost no measurable precipitation in a typical year. Yet thanks to massive ocean upwelling currents just offshore, the almost barren terrestrial environment is juxtaposed with one of the most productive marine ecosystems in the world. Indeed, it today supports what in good years is the world's most productive fishery—one based almost entirely on the production of a single marine species—the anchovy.

Back then, however, the chief resource was the shellfish that could be gathered close to shore, not the fish that teemed in the waters just offshore. Eventually though, as these societies continued to grow, they began to experience severe and periodic depletions of these sedentary marine resources. These crises began around 1000 B.C. and probably unfolded in the following sequence: first, a climatological catastrophe such as the El Niño phenomenon caused a sudden and widespread reduction in marine resource supplies along the coast; second, the large, sedentary, and concentrated populace that had been accustomed to relying heavily on these resources probably attempted to exploit them at their usual levels, thereby preventing their recovery; and finally, faced with the collapse of their fishery, these coastal peoples had no choice but to develop new means of subsistence in order to survive, developing "occupational pluralism" (which, as we will see in a later chapter, is a passive strategy of modern fisheries management).

It is possible these peoples also eventually perceived the need to institute active means for reducing and controlling fishing effort in order to address the crisis in their fisheries. Since they had already developed politically to the level of complex chiefdoms or nascent states, they had the kind of social organization required to implement "top down," centralized control over their fisheries. And while any fisheries-management systems they may have developed remain unknown to us, it is still reasonable to assume that they developed some, since practically every similar society for which we do have good historical and ethnographic evidence has done so.

Michael Edward Moseley (1975) contends that it is here that the early springboards of Andean civilization are to be found, not in the high Andes, as is commonly assumed. He describes how the severe depletions of the marine resources on which these societies had traditionally depended forced them to develop complementary means of subsistence, particularly agriculture along the river courses that cut through the otherwise arid coastal plain. This resulted in the development of more diversified and resilient forms of social and economic organization, as well as more centralized forms of political organization, paving the way for the powerful Incan civilization that subsequently arose in the Andean altiplano.

Two eminent archaeologists, Michael D. Coe and Kent V. Flannery (1967), describe the thriving of similarly dense and sedentary human settlements farther north along the Pacific Coast, in present-day Guatemala, between 1500 and 1000 B.C. As in Peru, these maritime societies may have provided the springboards for the florescence of the civilization found shortly thereafter in Mexico's high and fertile inland valleys. "The number of Early Formative villages per square kilometer in the lagoon-estuary system of the Guatemalan Pacific Coast," Coe and Flannery state (p. 5), "would seem to far exceed the number of permanent villages per square kilometer in the Tehuacán Valley. . . . Moreover, the year round stability of the coastal settlements appears to have been greater."

While these early coastal Guatemalans hunted and gathered food plants on the land and practiced a rudimentary form of agriculture, the archaeological record suggests they relied primarily on marine resources, particularly shellfish and the fish harvested from the surrounding estuaries and lagoons. Again, given their level of social, political, and economic development, they probably had the necessary cultural organization for controlling fishing effort and allocating its production. Remarkably also, there is some evidence that they traded and communicated with their contemporaries living much farther south, along the Peruvian coast (Coe and Flannery 1967: 4).

Guatemala's large neighbor Mexico, which is especially proud of its heritage from the ancient civilizations that flourished in its high valleys, has almost ignored the role maritime societies might have played in promoting the development of those civilizations. And this is very curious, considering that Mexico's first high civilization, the Olmec, arose not in a high valley but in a coastal zone. Much as in the case of the early coastal Peruvians, huge shell mounds scattered throughout the coastal zone attest to the Olmecs' high reliance on sedentary marine organisms.

The Olmec civilization sprang up around 800 B.C., along the Gulf coast in what are today the states of Tabasco and Veracruz. (Nowadays this seems an unlikely place for a civilization to have arisen, since it is a rather inhospitable place for human habitation. The climate is hot and humid practically the year round, and there are scores of biting insects. The surrounding terrain is mainly impenetrable marshes, labyrinthine systems of estuaries, and countless briny lagoons. But the region may have had a milder climate and a bit more dry land when the Olmec civilization first arose.) The Olmec built monumental works of architecture and art, had complex religious-belief systems, and in every respect seem to have been organized at the level of nascent states.

A few scholars have suggested that all this was prompted by culture contact with early mariners from the advanced chiefdoms and nascent states of West Africa (see, e.g., Heyerdahl 1979: chs. 3, 14; Schwerin 1970).

As they would have it, African coastal raft fishermen could have drifted offshore and along with the ocean currents leading to the Mesoamerican coasts, a voyage, they say, that would have required no more than four or five months at sea. This, of course, is pure conjecture, and while modern-day mariners like Thor Heyerdahl have proved such a crossing could have been done, it remains to be shown whether there was ever even one ocean crossing, from anywhere, prior to the Olmec florescence. This view is refuted by most Mesoamerican archaeologists, who cite the scarce and anecdotal nature of the evidence.[2]

Still, we must consider that the material record could have been impoverished by the extraordinary destructive forces at work in the Olmecs' coastal environment, and even if the Olmec had been contacted by African mariners (or other distant ones), there is little likelihood that the artifacts of these meetings would still be preserved. After all, it has been over 2,000 years since their civilization was at its peak.[3] What is certain, in any event, is that this civilization's development was supported by a high degree of reliance on marine resources. Indeed, the largest material deposits

[2] A different view concerning the possibility that the Olmec were contacted by transoceanic mariners appears in Jett (1983: 350–54), who summarizes works suggesting the Olmec were contacted by the Shang, the first Bronze Age civilization in the Yellow River valley area of North China. Other possible transpacific contacts with precolumbian societies are discussed in Jennings, ed. (1983); Meggars (1975); and Meggars and Evans (1974). Both Ferdon (1963) and Heyerdahl (1963) discuss the most probable routes of early drift-voyage mariners.

[3] I am thinking especially of fishing-camp sites I have visited in both Pacific and Gulf coastal Mexico that thrived in the 1930's. Those sites are already very poor in material remains, even though several hundred people worked at them a little more than fifty years ago. Because of the destructive forces of the sea along coastal shorelines, marine archaeology has focused inordinately on submerged shipwrecks rather than coastal settlements, which are far quicker to disappear over time.

of Olmec culture surviving today comprise neither their monumental works nor the portable artifacts they left behind, but rather the many shell mounds they left scattered throughout the coastal zone. These indicate that a large number of organized laborers engaged in both extensive and intensive fishing over a long period of time, and here again it is likely that the chiefs or kings controlled both the fishing effort and the distribution of marine production.

North of Mexico, along the coast of California, there is similar evidence that many of the early coastal peoples regulated their fishing activities. McEvoy (1986: 21) for instance, in his book on California's fisheries, comes to the following conclusions based on his reading of various archaeological studies: "Prehistoric fishing economies had to adapt in their own ways to problems that have plagued modern industries exploiting the same resources at the same localities. . . . We can infer . . . from what record of their activities we do have how Indian communities eventually learned to balance their harvest of fish with their environment's capacity to yield them."

Precisely how these early maritime societies managed their fisheries, if indeed they did manage them, we may never know for sure, although we can get a pretty good idea by studying similar societies that are well documented in various historical and ethnographic accounts. Moreover, whether or not they did take steps toward fisheries management, they almost certainly were challenged to do so. But what about societies that apparently did nothing to meet this challenge?

Unregulated Fisheries in Prehistoric Times

Unregulated fisheries are those in which fishing peoples apparently do (or did) nothing to limit fishing effort and reduce the mortality of the main marine resources on which they depend. We may also include here the few instances in which the fishing peoples seem to have flouted regulatory controls imposed on them by external agencies or entities.

The historical record presents us with a checkered array of case studies; a few portray fisheries that seem none the worse for their society's permissiveness, but most illustrate the dire consequences of unbridled effort. Most fisheries that failed to sustain accustomed yields are ones where the local culture apparently had an ineffective or nonexistent management policy.

Apparently had an ineffective or nonexistent management policy, I say, because from the record we cannot always discern exactly what led to a fishery's failure. Sometimes all we can do is speculate that the fishery either suffered some sort of ecological disaster or declined because of unbridled effort, or both. Whatever the case, even if the local people did employ some means of management, those means, appraised post hoc, cannot have been effective.

It has often been asserted that long ago, before the advent of modern technology and massive population growth, early peoples lived in harmonic equilibrium with their vital food resources. This natural equilibrium is assumed to have been a primal, almost intuitive means by which human cultures maintained a balance with nature. And as we look backward through the long telescope of time, it does seem as though many ancient cultural systems enjoyed a considerable run of long-term stability. Their subsistence practices, which are often characterized as reflecting a high degree of reverence for nature and a natural tendency toward self-restraint, were positively adaptive for them. Thus, the argument goes, this reverence and restraint became instrumental aspects of their cultures, and had much time to simmer, steep, and become institutionalized.

This basic harmony with nature—which is practically a cliché in the literature concerned with preindustrial peoples—is also described as having been congruent with the main ideological orientations of these preindustrial peoples, with their "bioethic" or their ethos regarding the natural world. It is an orientation said to bespeak a reverential attitude toward the natural world, an understanding of how the needs of the individual had to be merged with those of society and of how society's needs in turn could not be allowed to undermine the natural resources on which it depended. In this view, then, preindustrial peoples were "enlightened predators."

➤ Recent studies have challenged such thinking as naïve and simplistic. In contemporary conservation theory, considerable controversy surrounds the meaning of such concepts as "indigenous resource management," "enlightened predators," "optimal foragers," and "efficient harvesters" (see, among others, Chagnon and Hames 1979; Charnov 1973, 1976a, b; Gross 1975; Hames 1987; Hames, ed., 1980; A. Johnson 1982; Krebs 1978; Krebs and Davies, eds. 1978; Krech, ed. 1981; McCay 1981b; McDonald 1977; Pyke et al. 1976; Ross 1978; Stocks 1987; Vickers 1980; Winterhalder 1983). Raymond Hames (1987: 92), for one, maintains that "there have been few attempts to demonstrate the behavioral correlates and environmental impacts of cultural practices believed to have conservation functions." That native peoples had ideologies concerning potential resource availabilities, he stresses, does not prove that they were implementing strategies for conservation. Moreover, it also does not mean that their putatively conservationist ideologies constituted effective strategies for management. He joins many others who feel there has been a widespread tendency to romanticize simple societies by concluding that their members were conservationists merely because the resources they relied on remained abundantly available for long periods of time, a proposition with which I am in fundamental agreement.

However, I cannot subscribe to Hames's further assertion (1987: 93) that "any persuasive account of conservation as a human adaptation requires a theory that shows that conservation is by design . . . and not a side-effect of some other process." To the contrary, it is often the case that the members of local societies are unconscious of the incremental, trial-and-error adaptations that eventually became instituted as core beliefs and customary behaviors. Moreover, once positively adaptive culture traits and customs have been instituted, they may persist for a long time without any conscious human intent to make them persist.

Cultures develop as ongoing experiments. Thus there was nothing inherent in preindustrial cultures assuring that they would live in a state of harmonious equilibrium with nature. Instead, as a growing number of documented cases suggest, many early peoples wantonly wasted and depleted their important food resources, after which they may have learned some bitter lessons. Those who did not risked cultural extinction, meaning their particular experiment was concluded, or at least were forced to move on, to exploit other resources elsewhere. But the groups that survived and continued to harvest the resources they had earlier depleted must have developed new cultural institutions that reflected the lessons learned from past abuses. It little matters whether such groups incorporated what they had learned by resort to the patterns of thought and conscious design that we associate with modern conservationist practices or relied on other modes of thinking peculiar to themselves. All that really matters is that they internalized these lessons somehow.

And of course most of the earliest human societies were "enlightened" only because they had neither the numbers nor the technology that would have enabled them to deplete all their food resources. So these peoples' harmonious coexistence with nature seems to have been essentially a consequence of their simple inability to overtax their important ecosystems, whatever their view of their overall place in the scheme of nature.

Another commonly encountered myth about early humanity is that it coevolved with many of the organisms on which it preyed, culling the "unfit" and so ever-improving and sustaining the vigor and vitality of the prey populations it depended on. That assumption, however, has little validity in the case of early maritime peoples. Richard C. Hennemuth (1979: 8), for example, doubts "that man can be a prudent predator [in the marine environment], taking only what would die or not be produced in his absence." Even if we accept the contention that terrestrial hunters typically harvested mainly weak, sickly, slow, or aged creatures, weeding out the unfit to the benefit of the species' evolutionary development, the theory does not apply in the case of marine resources. Early maritime peoples largely confined their fishing to the waters along the shorelines and took what they could; they were hardly in a position to pick and choose.

There is an accumulating body of archaeological evidence showing that in certain microenvironments Pleistocene-era peoples did in fact sometimes overharvest marine food resources. David R. Yesner (1980), for example, cites instances of disastrous overharvesting in the New Zealand area, as described in the archaeological accounts of Wilfred Shawcross (1975) and P. Swadling (1976); in parts of Oceania, as reported by Patrick V. Kirch (1979) and F. M. Reinman (1967); and in Alaska, based on Yesner's (1977) own studies.

Generally speaking, these cases of early prehistoric marine-resource depletion involved hard-shelled, sedentary marine resources such as clams and oysters; there is not much evidence for the depletion of more mobile marine resources such as shrimp, fishes, or marine mammals. But there are a few notable exceptions. Charles A. Simenstad, James A. Estes, and Karl W. Kenyon (1978), for instance, describe how aboriginal hunters nearly eliminated the sea otter from the Aleutian archipelago. Still, again, the perishability of the remains of these creatures, as against the hard-shelled ones, may cause them to be underrepresented in the archaeological record.

In any case, it would be erroneous to assume that the earliest prehistoric maritime societies had an insignificant impact on marine resources. Shawcross (1975: 62), who has long studied the impact of prehistoric societies on marine resources, concludes that "the scale of the impact of prehistoric man on animal populations has been small in comparison to that of modern man, but it has a long history and cannot have been negligible."

Similarly erroneous, therefore, are suggestions that the marine species targeted by early prehistoric societies constituted "virgin populations," an assertion we occasionally hear from marine biologists concerned with conducting baseline studies of marine life.

One example of overfishing in more recent prehistoric times has already been discussed: the depletion of shellfish by the maritime peoples who lived along the Peruvian coast nearly 3,000 years ago. Similarly, and farther north along what is now California's Pacific coast, Clement W. Meighan (1959) describes aboriginal communities that disappeared after depleting their supplies of shellfish. Moreover, severe depletions were not exclusive to just shellfishers. McEvoy (1986: 19–62), for instance, provides numerous examples of California's aboriginal peoples suffering severe reversals in their salmon fisheries as a result of unbridled harvesting. Citing a variety of studies, he concludes that repeated boom-and-bust cycles occurred in many of California's aboriginal salmon fisheries long before European contact—often at the same sites where boom-and-bust cycles developed in modern times.

Unregulated Fisheries Prior to Modernization

R. E. Johannes (1978), in an important work concerned with traditional methods of marine resource conservation in Oceania, records a number of instances where unrestrained harvesting by indigenous peoples led to the depletion of important subsistence resources. And not all of these resources were as immobile and territorially fixed as clams and oysters. In the Solomon Islands, for example, porpoises were harvested for their teeth while the valuable meat was left to rot in the sun, eventually leading to a shortage of that valuable food resource. In many parts of Oceania, poison was used to bring in the catch, a process that indiscriminately kills both fry and eating-size fish. Both the Trobriand and the Hawaiian islanders wastefully overharvested important marine resources, bringing about their near depletion and a scarcity of local food supplies. The natives of Satawal, in the Caroline Islands of Micronesia, overharvested turtle eggs, depleting that resource, and "horrible waste was sometimes committed by the Tahitian royalty at their feasts" (Johannes 1978: 355).

Johannes is quick to add that these depredations were exceptions to a more general pattern of wise resource management seen throughout most of precolonial Oceania. Maybe so. But exceptional or not, these observations document the occurrence of unbridled and unregulated fishing activity among premodern maritime peoples, suggesting behavior that is decidedly the opposite of what we would expect from an enlightened predator. Like modern humanity, some premodern maritime peoples demonstrated a capacity for wastefulness and greed.

We also have a fair number of cases of premodern peoples who apparently made little effort to regulate their fishing effort and yet did not suffer any ill effects. We know from the very comprehensive account of Raymond Firth (1965), for example, that in the primitive Polynesian economy, where fisheries were common property and fishing was unconstrained, the peoples survived without experiencing either resource depletion or severe internal competition. Similarly, Thomas Gladwin (1970), in his description of the people of Puluwat, Micronesia, makes no mention of any constraints on fishing effort. But for all of the rather rosy picture he paints of an open-access, unregulated, and highly productive subsistence-oriented fishery, we find that by his own account the Puluwatan population is small, the local marine resources are abundant, and the territory available for fishing activity is immense.

Likewise, until very recently the traditional peoples of the circumpolar Arctic regions seem not to have paid any penalty for fishing as freely as they liked. This can definitely be attributed to their small numbers, rudimentary technologies, and vast fishing territories, and need not be attributed to their aboriginal regard for conservation. In fact, now colonized and considerably acculturated but still comparatively small in numbers, these peoples, equipped with such modern technologies as high-powered rifles, snowmobiles, nylon nets, and motorboats, often demonstrate a blatant disregard for conserving important stocks.

Undoubtedly, some of the early ethnographic accounts of maritime peoples living freely off the ocean's bounty are idyllic, affronting one's intuition and common sense and making one wonder about the extent to which their authors romanticized the peoples they studied. In the foregoing cases I suspect that the main reason that the peoples studied suffered no deleterious effects from their uncontrolled fishing activities was their simple inability to deplete their key marine resources. In all likelihood their populations were small, their capture technologies rudimentary, their available fishing territories large, and their supply of marine resources abundant relative to demand.

Unregulated Fisheries in the Modern Day

Because modern examples of overfishing are so numerous as to be pattern rather than anecdote, let me cite just a few particularly striking cases. G. L. Kesteven (1976) describes how whole stocks have been wiped out by artisanal fishers using simple haul-seines in Indonesia and by weir and bag-net fishers taking migratory fish from the Tonle-Sap River and the Grand Lac in Cambodia. He also describes the nearly total depletion of SEA. migrating shrimp stocks by bag-net fishermen operating near the mouth of the Rokan River in Sumatra. Similar instances of overharvesting by artisanal fishers have been noted in Thailand, Malaysia, Singapore, and the Philippines, and among fishermen employing tuna traps around Sardinia.

None of these fishing peoples could be described as fully modern or industrialized. Rather, they are peasant or traditional peoples living within developing nations. Nevertheless, their zeal in overharvesting important marine resources seems to be congruent with the behavior of many fishers in the modern nations, who increasingly seem to be fishing as if there were no tomorrow.

Another widespread practice that has led to marine resource depletion is dynamite fishing, a method used by many small-scale fishers throughout the tropics and subtropics. This indiscriminate harvesting method is very wasteful; not only are other species killed along with the sought-after species, but many of the dead target fish are not recovered. Over time the practice also brings about the destruction of critical marine habitats, sometimes beyond any hope of productivity for years to come. Around the Philippine Islands, for example, before it was outlawed in the early 1970's, dynamite fishing, widely practiced from the beginning of the century, eventually had a disastrous effect on the small-scale fishers' highly productive, long-established corral-fishing sites (Spoehr 1980). Nowadays, more vigilant enforcement has mitigated the impact of dynamite fishing in most fisheries, but it is still a fairly common practice, though considerably more clandestine.

Access to modern technology is only one of the factors that have led fishers to engage in unbridled fishing effort. The small-scale peasant fishers Thailand of south Thailand were subjected to so much competitive pressure from large-scale, offshore commercial fishers that they soon found themselves in unrestrained competition amongst themselves. This heightened competition not only accelerated the scarcity of their main subsistence stocks (mackerel), but also prompted the breakdown of the traditional modes of social and economic organization that had once constrained the fishing effort (Fraser 1966). Another sort of problem beset the small-scale fishers of Lake Izabal, Guatemala. This common property, open-access fishery experienced a chaotic rise in harvest pressure as a result of a similarly chaotic rise in the population living around the shores of the lake. At the same time, more effective harvest technologies were introduced into the fishery, with the result that the catch for a given amount of effort declined substantially. Their response to the dwindling catch was "to fish more diligently and acquire more nets" (Dickinson 1974: 369). So unrestrained was the effort and acute the resource depletion by small-scale operators working the inshore fisheries of South Africa, in fact, that the central government was finally forced to impose regulations on the types and placement of gear and to mandate fishing seasons and area closures (Scott 1951).

These cases, I must emphasize, are not isolated examples but part of a worldwide pattern seen in many open-access fisheries today. Cordell (1978) notes that the marginalized small-scale fishers working in a common property, open-access fishery of inshore Brazil now take huge quantities of spawning fish as their main catches. Furthermore, because these people fish by certain tidal cycles, they all typically crowd into the same few fishing spots, leading to heavy pressure at these sites and to severe conflicts and competition over access to them.

Ignorance or indifference to past crises in their fisheries has also sometimes led modern-era fishers to engage in unbridled harvesting practices. This was the fate of California's salmon fisheries. As McEvoy (1986: 21–23) notes, many of the state's aboriginal Indians accumulated a store of information over the generations and were able to work out management systems that sustained high yields for centuries—yields that even approached those enjoyed by the immigrant fishers who displaced them. When these Indian cultures were driven to extinction as a result of European immigration during the late nineteenth century, much of their fisheries management knowledge was lost. But because the salmon stocks were then left alone for many years, when the new arrivals turned to the enterprise, they initially enjoyed harvests far greater than the fisheries' maximum sustainable yields. This created in the immigrants' minds an unrealistically high estimate of what their fisheries could produce, and the result was a series of boom-and-bust cycles. Nowadays, McEvoy notes, California's fisheries managers are attempting to retrieve what remnants of the aboriginal knowledge they can in an effort to protect these endangered stocks.

Of course, much of the uncontrolled fishing behavior described above can be seen as a reaction to some external deleterious factor: the encroachment of outsiders into traditional fishing territories, the intensifying of competition, the adoption of new and more effective technologies. But whatever the cause, we find that under the pressure of diminishing returns, fishers often increase their fishing effort, knowingly hastening resource depletion and abandoning whatever voluntary restraints they might have observed.

A number of researchers have proposed that there is a correlation between the breakdown of self-restraint among fishers and the process of modernization itself. Modern attitudes are different attitudes, they insist, and the market mentality provides no incentive for fishers in competition with others to restrain their efforts today so there may be some fish left for tomorrow. Thus the loss of conservationist ethics seems to parallel increases in market demand beyond what a given fishery can supply on a sustained basis. This phenomenon has been repeatedly seen among fishers who have turned from production for subsistence to production for cash to meet an enlarged market demand. The turtle-fishing Miskito Indians of eastern Honduras and Nicaragua are one well-documented example (Nietschmann 1974). Johannes's (1978) account of the demise of traditional systems for fisheries management with the colonization of Oceania is another. In both cases, a traditional subsistence economy was abandoned with the advent of economic systems emphasizing production for the market.

The characteristic lack of restraint that onetime subsistence fishers display when they begin producing for the market can be attributed to a fundamental difference in the goals inherent in the two systems. In traditional subsistence systems, where the main goal is to produce food, there is a finite and thus satisfiable demand for the product. But for people living in a market economy—whether rural or urban, traditional or modern, impoverished or affluent—there is no upper limit on the demand for cash. *[margin note: subsistence ~ market. ≟]*

When people produce fish to feed themselves, with perhaps a small surplus to barter or sell to their immediate neighbors, a point will be reached at which the aggregate demand is satisfied and further production will have no utility. But when people produce fish to sell for cash—cash needed not only for purchasing life's necessities, but also for realizing the myriad, often illusory dreams that cash-and-market systems can stimulate—then there may be no upper bound on the fishing effort. Whatever levels of production and income they achieve, there are always greater goals to aim for. More effective technologies may then be acquired, capitalization increased, and harvest levels pushed upward until the marine ecosystem gives under the strain. Indeed, even artisanal fishers who continue to employ primitive technologies may quickly deplete fish stocks if they begin producing mainly for cash. Access to a larger group of consumers can trigger them to increase their fishing capacity and move them closer to the possibility of depleting their fisheries (FAO 1983: 10).

Building a fishing boat in a shipyard in the harbor at Essaouira, south of Casablanca, Morocco. Photo by F. Mattioli, courtesy of the FAO.

A fisherman carrying a big fish to market at a fishing village north of Dakar, Senegal. Note the smaller fish dangling from the big fish's mouth. Photo by F. Mattioli, courtesy of the FAO.

Fresh fish being sorted out on arrival at a fish freezing and processing plant in Dakar. Photo by F. Mattioli, courtesy of the FAO.

A woman from a fishing village south of Dakar sprinkling salted water on fish drying in the sun. Photo by F. Mattioli, courtesy of the FAO.

Fishermen putting out to sea from a fishing village north of Dakar. Before the establishment of the 200-mile EEZs, most of West Africa's fish catch was caught by foreign vessels. Since then, the local countries' share has increased greatly. Photo by F. Mattioli, courtesy of the FAO.

Fisherwomen dragging nets in shallow coastal waters at Chocas in the province of Nampula, Mozambique. Photo by M. T. Palazzolo, courtesy of the FAO.

Women gutting and boning fish, Benin, West Africa. These women are members of a government-sponsored cooperative that processes and markets the fish caught by the men. Photo by T. Fenyas, courtesy of the FAO.

Elaborate fish barriers and weirs near Cotonou, Benin. Photo by W. Gartung, courtesy of the FAO.

A fishing boat in full sail off the southwest coast of Sri Lanka. Although lacking any means of propulsion other than their sails, these dugout outriggers can make good speed and are capable of ranging far offshore. Photo by M.-C. Comte, courtesy of the FAO.

Fishermen perched on poles embedded off the coast near Tangalla, Sri Lanka, waiting for the tide to bring in the day's catch. Photo by M.-C. Comte, courtesy of the FAO.

With the help of the Food and Agriculture Organization of the United Nations and the United Nations Development Programme, the Bangladesh government is surveying the fish resources of the Bay of Bengal, particularly those areas lying within its EEZ. The survey extends also to brackish waters and estuaries. Here workers on the project's research ship sort and carefully record the catch according to species, size, depth at which caught, location from shore, season, time of day, and so forth. Through its sponsorship of countless projects in developing nations around the world, the FAO has made an immense contribution to improving the welfare of fishing peoples. It has also been a world leader in improving fisheries management and marine conservation methods. Photo by I. Bara, courtesy of the FAO.

Fish strung on bamboo poles to dry in the sun, Bangladesh. Once dried, they will be as hard as wood and can be stacked in piles. Photo by I. Bara, courtesy of the FAO.

Fishermen from near Cochin, India, using a Chinese-type net called a *cherai* to catch crabs, prawns, and mullet from the Arabian Sea. These cantilevered fishing nets were introduced to India by traders from the court of Kublai Khan. Photo by I. de Borhegyi, courtesy of the FAO.

Mother and children fishing just offshore at Ha Long Bay, Vietnam. In the developing nations, and particularly throughout much of Asia, fishers such as these often provide practically the only animal protein available to the general populace. Photo courtesy of Michael H. Glantz.

Anchovies being sun dried on raised fiber-woven tables at the small fishing village of Mercedes in the Camarines Norte region, Philippines. Fish provides a substantial part of the Filipino diet, and fishing is one of the country's most important industries. Photo courtesy of the FAO.

A fisherman washing his fish to help keep them fresh for market, Camarines Norte region. In developing countries in the tropics, much of the fish catch may be lost through spoilage before it reaches consumers. Photo courtesy of the FAO.

Developmental Problems in Fisheries Management

J ust what is a "fishery"? For some it is a geographic location where fishing takes place: the "North Atlantic fishery," for example, or the "Georges Bank fishery." For others the term refers to the method by which fish are caught: the "ground trawl" versus the "pelagic purse-seine" fisheries. For still others, the term refers to a particular marine species—the "tuna fishery" or the "anchovy fishery." Often, these definitional attributes are combined, as in "Peru's pelagic anchovy fishery."[1]

At the most fundamental level, however, fisheries are a *human* phenomenon, since there can be no fishery without human fishing effort. Thus the economist Lee G. Anderson (1977: 22) defines a fishery as "a stock or stocks of fish and the enterprises that have the potential of exploiting them," with "enterprises" implicitly meaning human enterprises. Similarly, Alexander Spoehr (1980: 196) stresses that a fishery is "a socioeconomic technological system in interaction with a marine ecosystem." I am partial to definitions that underscore human involvement: Raoul Andersen's (1982 : 18), for example, which states that "fisheries are activities through which people link themselves with aquatic environments and renewable resources."

The Birth of Modern Fisheries Management

As late as the end of the nineteenth century, there was a general perception throughout the world that the supply of living resources in the oceans

---

[1] Hewes (1948) is one of the earliest works I know of that discusses the various meanings of the term fisheries. *Ground trawling* entails towing a net, or trawl, along the sea floor. It is still the main method employed in cold-water fisheries such as the North Atlantic and

and seas was for all practical purposes limitless. Thus the eminent British scientist Thomas Huxley could state in 1883: "I believe that the cod fishery, the herring fishery, the pilchard fishery, the mackerel fishery, and probably all the great sea-fisheries are inexhaustible; that it is to say that nothing we can do seriously affects the numbers of fish" (cited in Pontecorvo 1986: 6).

The residents of local fishing communities knew better, of course, and by this time many had already developed local or regional regimes for managing their fisheries. Unfortunately, these regimes were known only to them and recorded only in their oral and behavioral traditions. As a result, few of the nonfishing peoples of the literate world knew how vulnerable to overharvesting many marine fish stocks actually were.

As industrial growth began to seriously pollute many productive nearshore fisheries, undermining their productivity, many of these alarmed local peoples began to pressure their governments to protect their fisheries against developments that they had no ability to control themselves. McEvoy (1986: 100), for instance, notes that "governments in the United States became actively concerned with fisheries after the close of the Civil War, as more and more East Coast fisheries succumbed to overuse and industrial pollution."

The Industrial Revolution also quickly extended to offshore fisheries, particularly during its second phase beginning in the 1890's, as internal combustion engines and the widespread development of electric power brought even more effective technologies into the fisheries. Sail gave way to steam, then to gasoline and diesel engines, which were able to lug trawls far larger than those that could be towed previously under sail. No longer constrained by the whims of weather, wind, and currents, fishing vessels could go practically anywhere, fish farther offshore, and reap harvests in ocean regions never fished before. Thanks to the security of their larger ships, and eventually to the development of other equipment, such as on-

---

North Sea, as well as in shrimp fisheries in warmer waters. In *pelagic* or *midwater trawling*, the net or trawl is towed somewhere through the water column above the sea floor. Most pelagic fish, however, particularly those that swim near the surface, are caught not by midwater trawling but by *purse seining*. This method, in which schools are enclosed in a large net, or seine, is favored for harvesting the various members of the tuna family. Other highly productive methods used in fishing today include *longlining*, in which baited hooks are positioned at regular intervals along a line up to several kilometers long, which is then either deployed along the sea floor or floated near the surface; *trolling*, in which several lines with either baited hooks or artificial lures are towed behind a vessel; and *drift netting* in which floating gill nets are deployed in the open sea.

board refrigeration, these industrialized fishers could stay out for much longer periods of time than before. And on their return, they were served by increasingly sophisticated processing, marketing, and distribution systems. As the world's human population continued to rise exponentially, fish production soared.

This offshore boom in fish production was fairly short-lived, however. As the nineteenth century neared its end, scarcities of certain marine resources that had always been abundant began to be noted, particularly in the North Atlantic and North Sea fisheries. As early as 1893, just ten years after Huxley's sanguine statement, a Select Committee of the English House of Commons expressed alarm that both the size of the total catch and the size of the fish being caught were diminishing in these fisheries. Clearly, overfishing was the cause. The widespread assumption that the living resources of the oceans and seas existed in almost unlimited quantities began to be seriously questioned for the first time (M. E. Smith 1982: 61). The result was the birth of modern fisheries science and the modern practice of fisheries management.

Informal discussions began between the North American and European nations exploiting the North Atlantic and North Sea fisheries that eventually led to the first multilateral fisheries conferences. One of the first was convened in Stockholm in 1899, under the sponsorship of Sweden. It led to the formation of the International Council for the Exploration of the Sea (ICES) in 1902, still the main organization involved in determining overall management policies for the North Atlantic and North Sea fisheries (M. E. Smith 1982: 61).

Fisheries management was thus transformed from a mostly local, regional, or national concern to an international one. In the process it had also gone offshore. Most nations with significant fisheries and a fishing industry now readily acknowledged an urgent need to bring their fisheries under better management, not only those close to their shores but also those in distant waters. And with increasing government support, the number of fisheries management agencies, management professionals, and fisheries scientists grew apace.

Although both the First and the Second World War temporarily curtailed fishing, giving many fish stocks time to replenish themselves, the overall fishing effort increased dramatically throughout the first half of the twentieth century. Then, in midcentury, just when fisheries managers were beginning to get a good handle on management, considerably more effective technologies were introduced. During the 1950's and 1960's, many fisheries saw the introduction of far more effective nylon nets, outboard motors, hydraulic power blocks (which allowed fishers to deploy far larger purse seines than any used in the past), electronic fish-locating equipment, and a whole host of other sophisticated types of fishing gear.

Where overfishing had once been seen as problematic in only a few major high-seas fisheries, it was now seen as reaching crisis proportions in many of the world's major fisheries, including many close to shore. By the late 1960's and early 1970's, the total world catch leveled off. The world-wide boom in fish production that had begun with the Industrial Revolution was over, and in many regions fish production even began to decline.

Since that time, new crises have prompted new approaches to fisheries management, but overall these approaches are still heavily influenced by the legacies of the management policies and regimes that were formulated in response to the crises in the large-scale, industrialized fisheries in the North Atlantic and North Sea around the turn of the century.[2] As a result, many of these approaches are fundamentally inappropriate for the management of small-scale fishers.

Bioeconomic Modeling and Fisheries Management

It was widely assumed among the first scientists to consider fisheries management that overfishing in the North Atlantic and the North Sea was caused by taking spawning fish out of the ecosystem.[3] This view, known as the *propagation theory*, was elucidated by the English biologist E. W. Holt (1895). He insisted that fish should have a chance to spawn at least once before being caught and further proposed that fishing profits could be increased or at least maintained by artificially producing and releasing fertilized eggs into the sea. But his fellow biologist C. Petersen (1894) did not subscribe to his theory. Petersen contended that even in heavily fished populations, the reduction of spawning fish was not the main reason for the decline in the high-seas fishing industry's profits. Rather, the main

---

[2] Three excellent works concerned with more contemporary aspects of fishing and fisheries management in the North Atlantic and North Sea fisheries are Andersen, ed. (1979); Pontecorvo, ed. (1974); and Warner (1983).

[3] This brief historical review of the development of bioeconomic modeling theories in fisheries management draws mainly on Anderson (1977); Beddington and Rettig (1984); Bell (1978); and Waugh (1984). Also useful were Crutchfield, ed. (1959); Hamlisch, ed. (1962); and Turvey and Wiseman, eds. (1957).

problem was taking fish that were too young and too small. His thesis, which is known as the *growth theory*, was that overall profitability could be secured if fish were permitted to grow before capture, since in most cases larger fish are more desirable to human consumers and bring better prices per unit of weight in fish markets.

Petersen's growth theory, based as it was on economic as well as biological considerations, constituted a significant point of departure from the more traditional and orthodox propagation theory of Holt and his followers. It was greeted with skepticism by many fisheries scientists when it was first proposed, and it remained in limbo for many decades, until more sophisticated means for measuring fish production were developed. But today it is a cornerstone of fisheries policies for many of the species targeted by fishers. Holt's propagation theory figures importantly only in the management of a few species whose fecundity is particularly critical. For these species, the artificial propagation of fertilized eggs is still an important management strategy for maintaining catch levels. Nevertheless, the practical application of the growth theory has remained elusive. The precise determination of the optimum age for capture remains an unresolved problem.

Petersen's theory was subsequently refined and given greater theoretical precision by such men as F. Baranov (1918), a Russian fisheries scientist, who was among the first to attempt to formulate a comprehensive yield equation that would equate levels of fishing effort with such variables as natural recruitment, natural mortality, and the growth patterns of age classes in a fish population. Still later, E. Russell (1931) set the theory into formal mathematical form. His formulations were augmented by M. Graham (1935), and then thoroughly formalized by R. Beverton and S. Holt, who by 1957 possessed excellent data from a large number of age-distribution studies corresponding with different levels and types of fishing effort.

By the mid-1950's, Petersen's growth theory and its mathematical formalization by Russell (1931) had been validated by a large number of studies that related levels of fishing effort with the age distribution of the catch, and it continues to be refined and validated today (e.g., Gulland, ed. 1977). In essence, the aim of all these formulations is to determine equilibrium levels between a fishery's biological productivity, the level of fishing effort, and the rate of fish mortality.

An important contribution in this vein was made by Milner B. Schaefer (1954). Schaefer presented a rigorous mathematical framework for explaining the relation between fishing effort and catch. Using data comparing levels of catches and fishing effort taken from the yellowfin tuna fishery in the eastern tropical Pacific, Schaefer's formal framework, now termed the *logistic model*, estimated the proportion of the stock or biomass that would be removed by a single unit of fishing, while also taking into account the intrinsic ability of the stock to increase and the maximum size it could theoretically attain. These estimates could then be used to predict both the maximum average yield the stock could support on a sustained basis and the average yield for any level of fishing effort. Schaefer's formulation thus solved one of the most perplexing problems that had faced early fisheries managers: how to determine a fishery's maximum sustainable yield, or MSY.

MSY

Meanwhile, other scholars were attempting to study the economic implications of these theoretical findings. This concern reflected a growing acknowledgment in fisheries management circles that overfishing was as much an economic problem as a biological one (see Graham 1943; Schaefer 1957, 1959). The optimum level of permissible fishing effort, it was stressed, depended on the market value of the catch and therefore could not be determined by merely determining the level of effort that would bring about a fishery's MSY.

This idea was explored by several scholars who now figure prominently in the development of fisheries management theory. Graham, (1943), for example, argued that unrestricted fishing effort in an open-access fishery would lead to a dissipation of profit. As one of the first works to present in a formal way both the empirical evidence and the theoretical necessity for controlling fishing effort, his book, *The Fish Gate*, was a landmark in the development of fisheries management theory. Graham later published a more definitive statement of his views in the second edition of the book (1949). Graham's work was augmented and refined by Beverton (1953), who definitively showed the relationship between obtaining maximum profit and the yield curve relating catch and fishing effort, and by Martin D. Burkenroad (1953), who stressed that the relationship between market prices for catches and production costs was the most salient aspect of the overfishing problem.

The aim of these various economic studies was to develop an instrument for determining a fishery's maximum economic yield, or MEY. The development of the requisite mathematical frameworks thus required considerable ingenuity. A comprehensive framework would have to integrate considerations of what a fishery could produce—its biological productivity—with considerations of the effects of various levels of fishing effort on fish populations in response to the changing costs of fish production and the movements of fish markets that would ensue.

But however desirable in theory it may have seemed to manage a fishery for MEY, in practice this would prove an elusive goal. H. Scott Gordon, for example, in two articles (1953, 1954) that drew on formal economic theory, showed that the equilibrium level of fishing effort in an open-access fishery—a level at which fishers would merely break even—would always exceed the level of effort required for MEY. Gordon was thus one of the first theoreticians to clearly describe the developmental process that often leads to a fishery's collapse.

That process can be summarized as follows. First, when fishers initially enter an open-access fishery, they experience high catches and high profits. Second, this attracts other fishers to the fishery, which in turn prompts the fishers who were already there to increase their investment in vessels, gear, and other capital items. Third, a point is eventually reached at which the fishery becomes overfished, as manifested in dwindling fish stocks, reduced catches for a given level of effort, and a climate of cutthroat competition among fishers who are intensifying their efforts to catch the dwindling stocks. Fourth, and finally, catch rates and profits fall to the point where at best most fishers can only break even. At this point, any further increase in effort will bring about losses and force some fishers to leave the fishery.

This fourth stage in a fishery's development is known as its *bioeconomic equilibrium*, and however benign that tag may sound, it implies a disastrous situation indeed. As John R. Beddington and R. Bruce Rettig (1984: 1) describe it, resorting to the jargon of economics, once catch rates and profits are reduced to a level where fishers are just breaking even "the economic rent (value of landings minus costs of catching and delivering the fish) is completely dissipated." At that point, the fishery is being exploited by too many fishers who are now too heavily capitalized for their average catch rates. Moreover, unless there is a reduction in fishing effort, the fishery will remain indefinitely at this low equilibrium point and will be unable to recover—either biologically or economically.[4]

Here, then, is the sticking point: any fishing effort beyond the level that would produce MEY for all participants results in overexploitation; and, for

---

[4] This ultimate stage of development, which Garrett Hardin so poignantly calls "The Tragedy of the Commons" (1968), is fleshed out in Christy and Scott (1965). The theoretical frameworks and basic assumptions of this model for explaining the "tragedy" are critically examined in the next chapter, where we will see that it has several important flaws.

economists, anything short of a fishing effort that produces MEY is lament-able, since in any other case the maximum economic rent is not being ex-tracted from the fishery.

The formal determinations of MSY and MEY were milestones in the de-velopment of fisheries management theory, but they have not provided a panacea for fisheries management in practice. Their main contribution is in informing policies aimed at conserving fish populations and maximizing overall economic efficiency. They offer little advice concerning what man-agement strategies will work best in actual practice. Moreover, these lofty ideals, with their suggestion of enlightened levels of fishing effort, have not often been attained in the year-by-year operations of most common property, open-access fisheries. Why?

Most fisheries managers acknowledge that the MSY goal is flawed be-cause of the simplistic assumption on which it rests—that is, the assump-tion that for a particular marine stock there is a level of fishing effort that can be sustained year after year, one in which the recruitment of new fish to the fishery will be neatly balanced by the overall catch. While this might be possible in a few cases, most marine stocks live in a complex and con-stantly changing environment that produces what sometimes seems to be random or chaotic fluctuations in their populations, the ultimate causes of which are often only dimly known (see Gleick 1987: 62–63, 80). Thus many marine stocks undergo boom-and-bust cycles that are entirely inde-pendent of fishing effort. Because marine environments do not exist in a steady state and MSY is predicated on biological equilibrium, for all prac-tical purposes there is little chance of ever achieving maximum yields in an ocean ecosystem year after year. As McEvoy (1986: 7) states, "Targeting a conservative yield from a real fishery is thus a problem in stochastic, or random-variable analysis—more like predicting the weather . . . than, say, the sustainable yield of guppies from a well-maintained aquarium."

If MSY is flawed by its inherent assumption that a sustainable equi-librium can be achieved, so is the MEY assumption that the level of fishing effort can be equated with a sustainable maximum economic return to fish-ers. The main problem, of course, is that the MEY concept assumes the existence of a stable market for fish, whereas in reality the market for fish is often quite variable and influenced by the fluctuations of markets in many other sectors of the economy.

Moreover, even if the fish market could somehow be stabilized, in mod-ern, open-access fisheries, fishers feel compelled to increase their effort to the point where they are just breaking even. As Geoffrey Waugh (1984:

19−21) notes, "Fishermen, in the absence of secure property rights, look to average rather than marginal productivity." What this means in practical terms is that fishers working in a highly competitive open-access fishery will usually feel compelled to engage in a level of effort beyond that corresponding with MEY. Consequently, unless stringent measures are taken to prohibit fishing effort beyond the point of MEY, open-access fisheries almost inevitably become overexploited as their economic rent is absorbed by excess fishing effort.

Bioeconomic modeling, which began with the works of Holt and Petersen, came of age with the works of Gordon and Schaefer, and today the _Gordon-Schaefer model_ is practically an icon in fisheries management theory. But though this model has proved very useful for managing most groundfish and pelagic species such as tuna, it was not until the late 1960's that similarly useful models were developed for coastal upwelling fisheries. Much of the practice of fisheries management is now predicated on the presumed utility of these various models. But in some ways it is also confounded by them because of their tendency to oversimplify the complex dynamics of marine ecosystems and human behavior, and the interactions of the two. In short, many of these models assert ideals of how fisheries management _ought_ to proceed that are often at considerable variance with how it actually proceeds.

The Trouble With Economics

When fisheries managers extended their attention beyond the biological aspects of managing the fisheries, it was natural that the discipline of economics captured their attention. After all, Maiolo and Orbach note:

Economics has a language, a method and, for many, a track record that are more impressive to the nonsocial scientist than sociology, anthropology or other social science disciplines. [But] the formal models such as those used by economists generally run into difficulty. Although they embody appealing language and method [they] somehow fail to predict the necessary range of consequences of policies. Indeed, in some cases . . . behavior proceeds in a direction opposite to that predicted! Such seems the case in many areas of fishery policy (1982: 13).

Similarly, Bailey (1988b: 113) observes, "The concept of MSY has nothing to say about the allocation of access to the resource itself," while "MEY as a concept provides no guidance regarding how resource rents should be distributed through society." These concepts have "the appearance of rationality," he notes, whereas in reality they "provide a convenient fig leaf of objectivity to cover what is in essence a highly political process." In other words, the models often have only limited usefulness when it comes to deciding crucial policy issues *equitably*.

Fishers also play an important role in confounding the usefulness of bioeconomic models by not behaving "rationally," at least not as rationality is defined by many economists. Because of their commitment to fishing as a way of life, they will often continue to work in a fishery even in the face of diminished stocks, dwindling yields, and extremely substandard incomes—something that drives economists crazy, since, from their point of view, this is "irrational" behavior. According to those economic theories that posit perfect competition in a world of perfect information, fishers would merely seek more remunerative means of making a living elsewhere once their fisheries no longer provided satisfying returns.

To the contrary, the tenacious behavior of fishers in the face of economic marginalization appears more rational when we examine real people who live in the real—and not theoretical—world. It is altogether understandable that many fishers will persist in working in a failed fishery when we consider that they are often poorly educated and ill-prepared for jobs that do not involve skills learned in fishing. Indeed, even well-educated and highly mobile professional people will often staunchly resist taking leave of a way of life to which they have long been accustomed.[5]

Despite the fact that the real behavior of many fishers does not conform with what economic models predict, a majority of fisheries experts persist in relying on them as benchmarks for establishing fisheries policies and regulations. After all, bioeconomic models are still the best theories they have, and the further development of this body of theory will clearly be important for improving fisheries management policies. But it is also clear that much greater account must be taken of other, less purely theoretical considerations.

The major problems in the fisheries today are not the biological depletion of fish stocks, economic marginality, overcapitalization, and so forth. Rather, they are the deleterious consequences of these conditions for the human participants in a fishery. For a fishery is not merely a marine eco-

---

[5] Orbach (1977: 111) offers another interesting reason for the persistence of fishers even in the face of declining stocks: their inherent optimism. Among the tuna seiners he studied, who indeed fish for a common property resource that they know to be in increasingly short supply, he found that "*on the most abstract level* there is very little of the if-I-get-it-you-won't, zero sum thinking." Instead, most of these fishers apparently assumed that the stocks might increase in the future regardless of the level of fishing effort.

system interacting with a human economic system; it is a marine eco-system interacting with human beings per se.

The prominent fisheries management theorist Geoffrey Waugh (1984: xx) would be seconded by many of his colleagues in his insistence that "successful management of these resources needs to be based on successful modelling. Theoretical modelling of the fishery requires the integration of the biological characteristics of the fish population with the economic theory of the firm." But successful theoretical models cannot be this limited. If they are ever to be truly effective, they must somehow integrate the biological characteristics of marine resources, economic theories of the firm, *and* considerations of human cultural traits, human needs and preferences, basic human nature, and human behavior in maritime settings.

The scope of a fishery manager's goal will be too narrowly defined if it merely emphasizes the maximization of the net economic return, because for many people the net economic return may be less important than some other, more desirable social goal—full employment, for instance. Of course, many economists would still argue that any condition of full employment that produces a less-than-optimal economic yield is inefficient and would over the long run provide fewer net benefits to all the participants in an economic system.

The error in this sort of thinking is that it does not acknowledge that much of humanity lived quite contentedly in the past in conditions that would be appraised as horribly inefficient today, and that even now there are many people who would not want to live and work in an economy whose chief concern was efficiency. The kind of fishing proposed as "efficient" by many economists relies on arbitrarily chosen variables, such as an ideal amount of time or labor committed to the effort, the employment of the best available hardware, and the garnering of the maximum economic rent. On that basis, fishing as it was conducted a hundred years ago is seen as inefficient, even though it was not inefficient at the time, and even though it might teach us much about how to safeguard fishing in the future. Similarly, "efficient" fishing today may seem inefficient a few years from now and may even be deemed counterproductive by future generations, who will undoubtedly perceive the world differently. The main problem, in short, is that economic efficiency and social efficiency are two distinct and not completely interrelated concepts.

Moreover, many of the economic realities that economists talk about are actually abstractions, and many economic theories concerning how economic systems function are even further removed from the real behavior of real-life human beings. Even the definitions of "economy" and "economic" are still being debated among orthodox economists.

A recurring problem inherent in many of the formal models economists employ is an implicit assumption that human behavior is everywhere pretty much the same, and that all economic behavior is motivated by a basic necessity in all societies to allocate scarce means to alternative ends in the most efficient manner possible. But this idea would find little support among cultural anthropologists. If cultural anthropology has contributed anything to our understanding of human nature, it has been to reveal the great diversity of human values, behavior, and social and cultural organization throughout the world. Thus, particularly in small, simple societies, the allocation of goods and services to alternative ends is often motivated by many factors other than those that are classically recognized as "economic"—kinship obligations, for instance, or the necessity of reinforcing alliances. As the substantivist Karl Polanyi (1958: 249) reminds us: "In the absence of any indication of societal conditions from which the motives of the individual spring, there would be little . . . to sustain the interdependence of the movements and their recurrence on which the unity and the stability of the process depends." So far as fisheries management is concerned, the substantivists' insistence that we must understand the societal conditions from which the motives and behavior of individuals spring is just as applicable to modern economic systems as it is to preindustrial or more primitive ones. For all economic systems are ultimately vivified and instituted by human actors who live in particular cultures with particular norms, institutions, and symbolic meanings regarding economic exchange (Kapferer, ed. 1976).

Thus fishing peoples like the Puluwatans of Micronesia, for instance, expend considerable amounts of time, energy, and resources constructing and sailing their oceangoing canoes. Sometimes their voyages are to distant islands where they visit friends and distant kinsmen, and sometimes they sail long distances in order to harvest sea turtles, even though there is an abundance of other high-quality marine seafoods immediately around their home island. From a modern economic point of view, these long voyages and the expenditures of human labor and natural resources they require are an affront to the ideal of maximizing net economic returns. For the Puluwatans, however, these voyages are motivated not so much by "economic" imperatives as by the need to reaffirm core attributes of their cultural identity—to demonstrate an expertise in navigation, for example, and to maintain their history as great seafarers (see T. Gladwin 1970).

In short, even with better formal models in hand, certain fisheries management decisions will probably never be determined by resort to mathematical frameworks alone. Allocating users' rights in a fishery, for instance, with the concomitant implications for social equity, is perhaps the most difficult decision fisheries management professionals have to make, and though the availability of better socioeconomic models may greatly inform such decisions, it is doubtful they will ever be made on the basis of formal models alone.

Fishers, Scientists, Managers, and Bureaucrats

Fishers, scientists, and fisheries managers share similar concerns for the conservation of fish stocks and for the prosperity of the fishing industry. However, their differences in social background, education, and enculturation, as well as their specific everyday concerns, often obscure this fact. Fishers in particular often express considerable antipathy toward the scientists and managers who regulate their fisheries. Bonnie J. McCay (1988: 329) describes the reaction of a friend and fisherman to the scientists and fisheries administrators he mingled with at a meeting. "He left the meeting disgusted at the 'objectivity' of scientists when men's lives are at stake, a not uncommon reaction of non-scientists to scientists."

At best, many fishers perceive fisheries scientists and managers as meddlesome people with whom they must cooperate; at worst, they see them as arrogant and insensitive bureaucrats who have the power to implement arbitrary and decidedly prejudicial fisheries regulations. Moreover, while environmental deterioration may in fact be the principal problem in certain fisheries, scientists and fisheries managers, unable to do much to control the environment, may inordinately focus instead on what they *can* control—the fishing effort.

Because the depredations of large-scale fishers prompted the first widespread public concern for restraining fishing effort, and also because these fishers manifested greater homogeneity than most small-scale fishers, they were fairly easy to identify and their catches were relatively easy to measure. As a result, they quickly became familiar entities among the early professional managers.

Little attention was paid to small-scale fishers, a far more numerous, more culturally heterogeneous, and more diffuse group. Moreover, not only was their aggregate impact on fish stocks considerably more difficult to assess, but their stress on individualism, freedom from restraint, and suspicion of strangers reinforced their estrangement from fisheries managers.[6] So in the early days of modern fisheries management, small-scale fishers were often little integrated into the management regimes, if at all.

Once the extent of the small-scale fishers' impact on resources became more widely known and increasing managerial effort was turned toward them, it was inevitable that many responded with resentment, antagonism, and resistance. A few fisheries managers exacerbated the problem by arrogantly exerting their authority or attempting to apply regulations they had learned in the management of large-scale fishing enterprises. When, in return, small-scale fishers, accustomed to fishing without much outside interference, reacted unfavorably and even unlawfully to the imposition of new regulations, they came to be perceived as confounders of the management regimes, and in many fisheries an ongoing antagonism arose between the two groups.

While it might have been better for the subsequent development of fisheries management if the early fisheries scientists and managers had first studied fishing peoples and small-scale fishers, rather than fish, they should not be faulted for their initial preoccupation with large-scale fishing. After all, it was the overfishing problems in the distant-water, industrialized fisheries, and not problems arising in small local fisheries, that first prompted widespread public concern for fisheries management. Besides, it was politically easier for public authorities to concentrate on these (mostly northern seas) fisheries, not just because the number of fishers impacted was relatively small, but also because the general public had become irate over the increasing cost and diminishing size of the fish that were coming to the marketplace.

On the other hand, even if the early fisheries managers had been initially more concerned with local small-scale fishers, they would not have found much in the way of published studies to inform their understanding of such peoples and their communities. Comprehensive studies of small-scale fishing peoples did not begin to appear until around the middle of the twentieth century, nearly 50 years after the crises in the North Atlantic and North Seas fisheries had come to public attention, with the publication of such landmark works as Raymond Firth's *Malay Fishermen: Their Peasant Economy* (1946), Edward Norbeck's *Takashima: A Japanese Fishing Community* (1954), and Thomas M. Fraser's *Rusembilan: A Malay Fishing Vil-*

---

[6] These remain among the most formidable problems facing fisheries managers overseeing small-scale fishers today.

lage in Southern Thailand (1960). So, while in hindsight it is apparent that modern fisheries management might have been better advised to pay more attention to the small-scale fishers from the start, in all fairness, critics must take account of the fact that fisheries management science was well along before public attention was drawn to overfishing problems in this domain.

Moreover, if there has been a continuing reluctance among fisheries managers to sponsor or undertake studies of local fishing peoples, it is for good reason. Mainly, it takes a lot of time and participant observation to gain a comprehensive understanding of such peoples, and there are also special problems entailed in the study of small-scale fishing cultures. As Pollnac (1984: 285) notes, while the social research methods used in inland and coastal zones are similar, certain characteristics of coastal fisheries present researchers with special problems that make their task more difficult. It is usually harder to visit the harvesting sites of fishers than it is to visit a farm or a factory, and the seasonal mobility of fishers often makes them difficult to locate. Many fishing communities are isolated and hard to reach, while going on fishing trips is time-consuming, if indeed the fishers do not object, and they frequently do, to having nonproductive people on board their vessels.

There are also inherent problems in most of the available studies of fishing peoples, which make them difficult to apply in the practice of modern fisheries management. Overall they show little uniformity in terms of their methods and conclusions, and apart from the various studies known as social-impact assessments, they rarely have much quantitative salience. For these reasons, many fisheries scientists and managers are skeptical about their potential utility for fisheries management.

Fisheries managers, whose first concerns have traditionally been the conservation of valuable marine resources, are naturally compelled to take a conservative approach to management, one that reduces the risks associated with the overall management policy to the greatest extent possible. According to Peter H. Fricke (1985: 47–48)), three main risks concern fisheries policymakers and managers: "adverse user and public comments, further deterioration of the resource, and challenges to agency competency." Taking his observations from a study by Mary Douglas and Aaron Wildavsky (1983), Fricke notes that "when there is a situation in which knowledge and experience are incomplete, and there is disagreement on the course to be taken, decision makers choose the course with the least risk."

Part of the reason why cultural anthropologists, sociologists, and other social analysts are not playing a larger role in fisheries policy and management, as Fricke notes, is the "organizational climate" in many fisheries management establishments. While many governments stress the use of cultural and sociological studies in setting resource management policy, few have followed through by establishing the organizational frameworks that would make this a reality. The important decisions are still being made by those from the traditional disciplines of biology and economics, and this is likely to remain the case until anthropologists and sociologists develop more rigorous methods of analysis in their studies of fishing peoples.

It is thus no surprise that policy makers and managers still rely to such a great degree on bioeconomic modeling formulations. However flawed these models may be in practice, they still seem considerably more reliable than any of the sociological models developed thus far. Yet by the same token, however flawed the social studies of fishing peoples, their value should not be overlooked entirely. They contain a wealth of insights about fishers that it would behoove fisheries scientists and managers to know more about.

Fisheries managers are not the only professionals who have had conflicts with small-scale fishers. Developmentalists have had their share as well. Presumably developmentalists are concerned about people and are committed to bringing about beneficial social and economic change. Unfortunately, many also appear to be mesmerized by certain images of the modern world and prone to emphasize certain technological changes without adequately considering their appropriateness in the targeted cultural milieu. Like their colleagues in fisheries management, most are university educated, employed in government, academic, or scientific institutions, and live in urban settings. In other words, they tend to be as far removed from the life and concerns of the small-scale fisher as fishery scientists and managers.

Well-intentioned but otherwise misguided developmentalists have sometimes promoted changes in fishing societies that have inadvertently brought about unemployment, overcapitalization, overfishing, and other forms of social and economic breakdown or strain, including the exacerbation of internal structural problems in those developing countries in which a small capitalistic elite controls the nation's politics and economy and prevents a more equitable distribution of the nation's production. Many of these problems might have been avoided or at least mitigated if local peoples had been integral participants in the planning, implementation, and monitoring of these development projects.[7]

Much as in the case of some fisheries scientists and managers, developmentalists have on the whole neglected to consider whether available social studies of fishing peoples might offer insights for formulating development policies. Many of the larger development agencies do now have at least a few cultural anthropologists or sociologists on their permanent staffs, to be sure, and many more are occasionally hired as consultants. Some of these agencies also employ expert fishers, either as permanent staff members or as consultants, in order to draw on their experience. But, generally speaking, both groups remain peripheral to the decision-making processes in these organizations, which continue to be dominated by specialists in the biological sciences, economics, engineering, and law.

Even in the United States, which passed federal legislation in the 1970's mandating consideration of social and cultural concerns in fisheries management, there has been no significant move to incorporate cultural anthropologists, sociologists, and expert fishers into the formulation of fisheries policy to any significant degree. As late as 1987, there was only one social scientist (and no expert fisher) among the 1,054 professional specialists permanently employed by the U.S. National Marine Fisheries Service. Biologists (785) and physical scientists like chemists and oceanographers (126) were in the vast majority (NMFS 1987).

In most fishing nations today there are few institutionalized forums where small-scale fishers can effectively express themselves, and fewer still where they are empowered to influence fisheries policies in any decisive way. Even in the United States, where fishers are represented on various regional management councils, they often come away feeling alienated. As E. Paul Durrenberger (1988: 211) observes, "fishermen fish for a living. They do not make a living by going to meetings." Moreover, even when the Gulf Coast shrimp fishers he studied do attempt to participate in the policy-making process, they can seldom exert much influence because "the management process rests on scientific rhetoric [and] the fishermen have least access to scientific work and personnel."

Indeed, significant advances in marine biology and fisheries management science in the era following the Second World War have considerably widened the information gap between what scientists and professional

---

[7] For a case study of such well-intentioned but uninformed efforts in Sri Lankan fisheries, see Alexander (1975).

managers now know about the dynamics of a fishery and what is known by most of the producers. As a result, some professional managers have been emboldened as never before. At several of the fisheries management council meetings I attended in the United States, I saw these differences between fishers and fisheries managers acted out in the following manner. First, some small-scale fisherman who is either a member of the council or has been authorized by his fellow fishers to speak for their interests steps self-consciously to the microphone, and begins by apologizing for his lack of public-speaking abilities and scientific knowledge as he faces a long table at which are seated several government officials, scientists, and academics. Then, after rambling for a few moments, the fisherman is interrupted by the Chair, who admonishes him, "Please, come to the point!" Frustrated and exasperated, the fisherman blurts out something like, "We who work on the water every day of our lives—year after year—we know more about this fishery and what needs to be done than you do. We've got families to feed, boats to pay off, weather to worry about. Why don't you just let us manage the fishery for ourselves?"

After the meeting adjourns for the day, informal meetings sometimes take place in various hotel rooms, and often it is here that the recommended changes in management policy are hammered out. Sometimes, depending on who is present, one overhears patronizing remarks about the fisherman who earlier testified—his misunderstanding of marine biology, his ignorance of the law, the narrowness of his point of view, and so forth. Especially when those present in the room are mainly from business, government, and scientific and academic institutions, the conversation may turn away from the complex human problems the fisherman referred to and focus instead on more abstract matters—what the Gordon-Schaefer curves show, for example, or how the ratio of productivity to the level of capitalization in the fleet is deteriorating.

The decisions these participants eventually reach may be predicated on what are still merely hypothetical ideas in fisheries science, but ideas that, unfortunately, have become so much a part of the council members' professional thinking and everyday parlance that they have taken on a false concreteness. Moreover, some of the professionals attending these meetings have hidden agendas that are never revealed, either in the public hearings or in the informal meetings behind closed doors. They may be representatives of vested interests whose main concerns are quite remote from those expressed by the fishers who speak up at the council meetings.

Eventually, these fishers may learn that the council recommended rules and policies counter to those they advocated, and it may not be good news that they take back to their constituents. They also may not fully understand the underlying reasons for these recommendations and may simply perceive them as working new hardships on all of them. Most will eventually come to feel that the time they spent away from fishing to attend the meeting was wasted, and this will increase their antagonism toward the policymakers in the future and further alienate them from the policymaking process.

M. Estellie Smith (1982) provides a more balanced view of the regional decision-making process than the dramatized view I present here. She concludes that the considerable confusion and difficulties encountered in establishing regulations and policies at these council meetings stem from problems inherent in the participant groups. Nearly all participants, whether they are fishers, processors, or administrators, reject fundamental propositions concerning fisheries needs and the already available scientific knowledge at hand on how to address them. Furthermore, many of them "present arguments so all inclusive and sweeping that they lack focus" or have no substantive basis, "fail to recognize the extent to which the industry itself has altered old attitudes, values, work and associational patterns relating to traditional fishing life," "defeat themselves by letting differences divide them," and "lack sufficient knowledge of parliamentary procedure" (pp. 88–89).

Particularly revealing is Smith's discussion of the conflicting hopes of the various groups in promoting the establishment of a 200-mile protected zone for U.S. fishers. The fishing industry saw the new law as a means of "kicking the foreigners out," whereas fisheries administrators and scientists saw it as an opportunity for conserving existing resources, that is, as a means for reducing overall fishing effort. "One group," she observes, "was thus concerned with exploitation and *growth in the fisheries*, and the other was aiming for *preservation of the stocks*" (p. 65).[8]

Maiolo and Orbach (1982: 2) describe the pervasive problem underlying the struggles between fishers, scientists, and government officials as the tendency for power to concentrate at the top and flow mainly from the top down, as well as the tendency of bureaucratic regimes over time to become

---

[8] My view of the council meetings is admittedly more simplistic and partial to fishers than Smith's. Nevertheless, her observations essentially support my impression that fishers often become frustrated when trying to make their views understood at these meetings.

very resistant to change. They observe that "while the aggregation of issues and information which constitutes the *formation* of policies has an *upward* thrust, the thrust of their *implementation* . . . is *downward*." Particularly with modernization there tends to be a shift from fisheries being managed locally and from the "bottom up" to their being managed bureaucratically, from a greater distance and from the "top down."[9]

Maiolo and Orbach further emphasize the self-generating and self-perpetuating tendencies of bureaucratic systems:

Once a precedent in law, policy . . . or administrative form is established, structural forces are set in motion that tend to perpetuate activity pursuant to the precedent. . . . Once a pattern of reasoning and behavior is established for authority or intervention in a large system with many players, it takes a tremendous amount of time and effort—in addition to the innovative effort required to develop and disseminate a *new* line of reasoning—to either strike or redirect that reasoning or activity (1982: 5).

In modern fisheries-management circles, where fishers' needs and desires are considered, the fishers themselves are rarely empowered to sway fisheries policies. Ideally, it is they who should make the ultimate decisions about the management of their fisheries, with the "experts" playing only advisory roles. But established orders and old legacies are slow to change, and few modern fisheries managers are willing to hand over the reins of power. Moreover, it is doubtful that they should, at least not until fishers gain a broader understanding of the problems that plague their fisheries and manifest a greater commitment to remedying them.

As I have said, we should not lay all the blame for fisheries management problems on the scientists and managers, since they are no less constrained than the fishers by immense economic, political, and legal forces not of their own making that can frustrate the best efforts to bring about constructive change. As McEvoy (1986: 183–84) points out, managing California's fisheries during the first half of the twentieth century was a bureaucratic nightmare because government had to wrestle with formidable problems of its own, including "interjurisdictional competition, inability to focus diffused or ill-defined social values, and the need to expend its own resources where they would have the most effect." Moreover, with only limited amounts of resources and time to devote to the management

---

[9] I am not certain who first applied the top down/bottom up distinction to fisheries management, but I am grateful to that person because it is a crucial distinction, underscoring as it does the origin and vector of power in fisheries management regimes today.

effort, and needing to show results, fishery administrators are often compelled to focus on immediate, short-term problems, distracting them from
considering the long-term trends in a fishery—in much the same manner
as fishers may be distracted.

One root cause underlying the difficulty of managing modern fisheries
undoubtedly stems from the inability of market systems to address social
and environmental concerns, particularly when common property resources—such as fish—are being transacted. As McEvoy (1986: 10) states,
"In a competitive economy, no market mechanism ordinarily exists to re-
ward individual forbearance in the use of shared resources." Furthermore:

The injuries that fishers impose on each other by overharvesting . . . or that water
polluters inflict on the fishing industry . . . do not normally come to account because the market diffuses them too broadly—too many victims each suffering a
little bit—for any one person or coherent group to demand compensation for them
in the marketplace or in the courts. They are . . . "social costs," that is, costs that
fall to society at large because the expense of making and enforcing contracts to pay
them is too great to make the effort worth anyone's while" (McEvoy 1986: 9).

Here McEvoy draws on James Willard Hurst (1982: 55–66, 131–41),
who underscores four inherent problems in modern market systems that
legal regimes and changes in public policy have always had great difficulty
addressing: (1) the usually restricted range of interests that actually drive
market actions and whose bargained transactions have ramifications in
much wider social and environmental contexts; (2) the limitations of a
money calculus that makes money—an abstract value—rather than the
market's effects on society and the environment the prime consideration
driving market transactions; (3) the market's capacity for ongoing but
nevertheless only limited, incremental change; and (4) "the market's ac-
ceptance or promotion of inequalities in the distribution of economic, social, and political power" (p. 55) by permitting private decision making
within the larger structure of power.

For McEvoy (1986: 119), the difficulty of minimizing the social costs
stemming from failures to bring fisheries problems under control in the
United States since the late nineteenth century can be laid to one fact—
that "the tendency of market forces to sunder the ecological bonds between natural resources and their environments, both cultural and natural,
proved more powerful than any incipient awareness of the steadily growing social costs of allowing them to do so."

Such are the pervasive problems and formidable constraints that everyone concerned with the fisheries—fishers, processors, distributors, consumers, scientists, managers, and politicians alike—find themselves embroiled in, and which they alone can hardly remedy. These problems are evolutionary in their development and global in scope, an integral part of the world's political economy, and just what their solution may be is beyond anyone's ability to say. Perhaps the best first step toward remedying them, then, is merely to increase a general awareness of them.

In essence, many fisheries problems are merely a small but interconnected part of more pervasive problems in the world political and economic order. Thus when we consider fisheries management more broadly, looking up from our usually narrow focus on a particular marine environment, the producers who work in it, and the managers who strive to regulate it, we see that it is actually an arena in which diverse societal, political, and market interests participate in an age-old struggle for the allocation and control of scarce resources—a struggle that is taking place within nations as well as among them. From this perspective, the ongoing struggle over fisheries resources is merely a microcosm of a much larger struggle in which diverse interests compete for the use of waters for fishing, industrial needs, waste disposal, recreation, transport, and so forth.

Redeveloping Fisheries Management

The early concern in fisheries science with the paradigms of biology and economics provided the cornerstone ideas and intellectual orientations from which much of today's modern management science has developed. These early concerns—with industrialized, distant-water fisheries, on the one hand, and biological and economic theories, on the other—also played a key role in fostering the development of a fisheries management subculture, which by now has its own theoretical ethos, as well as its particular customs, initiation rites, dialectical patterns for communicating, rules for group membership, rites of passage, and collective concerns for reducing its professional risks and ensuring its survival. This subculture also often perpetuates certain traditional approaches to fisheries management problems that use traditional clusters of certain types of professionals; unfortunately, these traditional clusters seldom include very many fishers, anthropologists, or sociologists.

This subculture could thus stand some redeveloping, and perhaps a good way to begin would be to reconsider various fisheries management theories. In the next chapter I will examine one of management's major paradigms, but first let us examine just one comparatively small aspect of current theory.

This has to do with regarding fishers as analogous to other natural predators on marine life. The underlying logic is that fishers, like all marine predators, contribute to fish mortality. Jeffrey Kassner (1988: 182), for example, examines New York bay fishers "from the perspective of a predator on hard clams requiring a certain amount of harvest to survive and as competitors with each other and with other predators for hard clams." His aim is to develop an ecosystemic model that integrates fishers and marine species into one analytic system so as to effect a more comprehensive means of fisheries management. This is a worthy aim so long as the analogy is not taken too literally. But I have to disagree with Kassner (1988: 182), as well as others who argue in a similar vein, when he states that "a fishermen population . . . is no different than that of any other species." To be sure, he implicitly means "no different" for purposes of the model he proposes, but even so, all other marine predators lack culture and are fixed into their patterned ways of life mainly by instincts. Seldom can they develop new or alternative means of subsistence, much less change their patterns of behavior in response to changes in a fishery.

Human "predation" is culturally learned and influenced, and is driven by many forces other than the need to secure food—by market demands, for instance, which are not "rational" in the same sense as the driving forces underlying the quest for food by nonhumans. One good example of these "irrational" (in terms of the rest of the animal kingdom) human practices is conspicuous consumption, which creates high market demands for prestige or fad marine food items. Another is the role aesthetic values play in determining what kinds of lifestyles and livelihoods are desirable. Still another is the uniquely human practice of fishing merely for fun. In other words, human marine predators do not fish solely for the purpose of feeding themselves, but go after fish for a multiplicity of reasons.

Equating humans with other marine predators is also flawed for another, more technical reason. Whereas human fishers extract the living matter they capture from the sea and transfer it to terrestrial ecosystems, most marine predators stay in the marine ecosystem and eventually return to it some of the biomass they have captured.

Therefore, though ecosystemic models characterizing human fishers as predators do seem to me to have considerable heuristic value for the development of more comprehensive and efficacious models for fisheries management—particularly those in which resource mortality is the main concern—the developers of such models should be careful not to extend the analogy too far by overlooking the complex and unique factors that motivate human behavior, many of which transcend the motivational bases underlying behavior in the rest of the animal kingdom. To do otherwise is simply to dehumanize human fishers.

FIVE

The Tragicomedy of the Commons

Among the major theories explaining why problems arise in certain fisheries, surely the best known is the one linked to the tragedy-of-the-commons model. It ascribes the root cause of problems in many fisheries to their status as common property and maintains that when tenure to marine resources is unspecified, a tragic situation will almost inevitably develop as more and more competitors enter a profitable fishery.

The theory constitutes a model for explaining overexploitation and overcapitalization that is easy to understand and thus very persuasive. Moreover, it is practically ubiquitous in discussions of the problems of fisheries because a majority of the world's sea tenure systems are indeed common property systems. Yet as we shall see, it is also a prime example of a theoretical misstep in fisheries science.

The Development of the Theory

The idea that the common ownership of resources is inherently problematic is not new. It was articulated by Aristotle some 2,300 years ago, when he said: "That which is common to the greatest number has the least care bestowed upon it" (cited in McCay and Acheson 1987: 2). The idea has been examined by scholars ever since. W. F. Lloyd (1833) spelled out the underlying general dynamic of common property situations in essentially the terms that characterize the idealized model today, and H. Scott Gordon (1954), as discussed in the previous chapter, spelled out the developmental process leading to a common property fishery's eventual collapse. But nobody did more to popularize the phenomenon than Garrett Hardin (1968) when he put the unforgettable name "The Tragedy of the

Commons" to it. Defining what he meant by "tragedy" (p. 1244), Hardin cited A. N. Whitehead's (1948) observation that "the essence of dramatic tragedy . . . resides in the solemnity of the remorseless working of things" (see also Hardin and Baden, eds. 1977; Wilson 1977). Hardin's model has since become so paradigmatic that it is, as R. K. Godwin and W. B. Shepard (1979: 265) note, "the dominant framework within which social scientists portray environmental and resource issues."

The tragedy of the commons that Hardin refers to is the presumably inevitable collapse that natural ecosystems, and the human economic enterprises that exploit them, will undergo when their key resources are common property, and the effort that is applied to producing them continues to increase. Acute competition, overcapitalization, and resource depletion are integral aspects of this tragedy. The model is thus predicated on certain assumptions about human nature, or at least human behavior under some specified conditions.

For Hardin (1968: 1244), herdsmen grazing their cattle in "a pasture open to all" were fitting exemplars of his theory. But because of the ubiquity with which marine fisheries are instituted as common property, it is hardly surprising that fishers have subsequently displaced herdsmen as the theory's main exemplars. What kind of behavior, then, should we expect of these exemplars? The theory suggests that once the number of competitors in a fishery is sufficient to bring about declining yields for a given amount of effort, individual fishers, rather than exercising restraint to conserve the stocks, will instead increase or intensify their efforts. Even as their margin of profit dwindles toward zero they will redouble their efforts, taking what fish they can today on the assumption that there will be even fewer fish to catch tomorrow. In this theory, though increasing effort in the face of declining yields seems "rational" from each fisher's point of view, such behavior is collectively ruinous—or "tragic"—for all the participants in the fishery. In Hardin's (1968: 1244) words, "The individual benefits as an individual from his ability to deny the truth even though society as a whole, of which he is a part, suffers."

Most fisheries management professionals have felt that the most practical way to prevent the tragedy in common property fisheries is to develop externally imposed, government-run management regimes that restrain fishing effort. Basically, they assume that fishers are unable to exercise self-restraint, not only by reason of the kind of myopic "rationality" described above, but for other reasons as well: their rugged individualism, their often low educational levels, their isolation and provincialism, and so forth. All of these common attributes of fishers, it is argued, leave them ill-equipped to manage their fisheries on their own.

Some professionals, particularly those who have become disenchanted with state-imposed regulatory regimes, have increasingly proposed a different strategy for preventing the tragedy: privatization (see DeGregori 1974; Demsetz 1967; Furubotn and Pejovich 1972; Peters 1987). The privatization of heretofore common property, open-access fisheries, they feel, would provide more compelling incentives for fishers to restrain fishing effort among themselves, as well as make them more responsible for bearing the cost associated with the managerial effort. These specialists also see the tragedy in common property fisheries as closely associated with "open access," which is assumed to be an attribute of most common property regimes. Thus the proponents of privatization believe the tragedy could be avoided if common property fisheries were converted to private property.

A. D. Scott (1955), for example, emphasizes that a sole owner who can collect an economic rent from a fishery is provided with a significant incentive to conserve its fish stocks for the future. Moreover, an owner who is not involved in ruthless competition with other producers will be prompted to harvest the resources of the fishery at comparatively low levels of capitalization and operating costs.

These two extremes—absolute government control of a common property fishery and absolute privatization—exemplify classic positions in a grand-scale philosophical debate concerning the best way to organize human societies and economies. Most professionals in fisheries management, however, propose solutions that fall somewhere in between. One typical middle-of-the-road position recommends managing a fishery as common property but limiting its use by conferring various types of access rights.

Still, a majority of fisheries scientists and management professionals, regardless of whether they are basically in the government-control or the privatization camp, take for granted the general validity of the tragedy-of-the-commons model, including its inherent assumptions about human behavior. Committed as these decision makers are to the model, it is hardly surprising that so many of them are skeptical about the possibility of developing fisheries management regimes that would permit user groups or local communities to manage their fisheries on their own.

Some Weaknesses of the Model

For all the popularity of the tragedy model—a popularity that, as McCay and Acheson (1987: 5) observe, "may be related to its ability to generate both liberal and conservative political solutions"—recent work in the social sciences, particularly in maritime anthropology, clearly indicates that it has many defects. McCay and Acheson (1987: 5), for instance, criticize the model for its assumption that open access usually goes hand in glove with common property, even though an increasing number of field studies have shown the contrary (e.g., Acheson, 1987, 1988a; Berkes 1985; Berkes, ed. 1989; Carrier 1987; McCay and Acheson, eds. 1987; National Research Council 1986; Ruddle and Akimichi, eds. 1984). Moreover, they contend (p. 34) that "by equating common property with open access, the tragedy-of-the-commons approach ignores important social institutions and their roles in managing the commons." Indeed, they note, when the approach to managing a commons ignores and supersedes local management approaches, instead emphasizing either government intervention or privatization, the change can substantially weaken or destroy local institutions that were effective in preventing the "tragedy" and may even encourage it (see also Berkes et al. 1989). Hardin's model assumes that the users of a common property resource can do little to change the system of exploitation themselves, but several recent studies of local fishing peoples have shown the contrary (see, e.g., Berkes 1977, 1987, 1989b; Cordell, ed. 1989; Klee, ed. 1980; Liebhardt 1986; Morauta et. al., eds. 1982; Ruddle and Johannes, eds. 1985).

These and other studies reveal that local fishers can and often do develop their own means of mitigating, if not avoiding, the tragedy of the commons, frequently by establishing cooperative institutions out of a mutual realization of their interdependence. Indeed, McCay and Acheson (1987: 15–16) conclude that if, as M. Estellie Smith (1984) would have it, comedy, the classical counterpart of tragedy, is a "drama of humans as social rather than private beings, a drama of social actions having a frankly corrective purpose," perhaps the growing documentation of local self-management in the fisheries suggests that there are many comedies of the commons, and not just tragedies.

In essence, what the growing number of studies of fishing peoples suggests is that the model as it is usually conceived is too abstract and generalized to provide much understanding of particular common property fisheries. While its elucidation undoubtedly had great heuristic value only a few decades ago, its subsequent reification without adequate refinement now threatens to undermine its general validity. "The commons," it turns out, does not mean the same thing in all fisheries, and the behavior of fishers does not everywhere conform with what the model predicts.

Hence Andrew P. Vayda (1988) and others have emphasized that, absent an understanding of local factors—those that restrain and allocate effort in particular—it is erroneous to assume that people harvesting common property resources must inevitably become involved in the tragedy of the commons. In other words, while etic analysis may identify resources as common property, it does not follow that this means the same thing emically, that is, from the point of view of the users. Such property's historical development, as well as social norms governing its current use, may differ considerably in different cultural contexts. Thus, as the eminent anthropologist Bronislaw Malinowski (1926: 20–21) observed some time ago in his studies of a simple society, ownership in a common property regime can only be understood with reference to "the concrete facts and conditions of use. It is the sum of duties, privileges and mutualities which bind the joint owners to the object and to each other."

Other defects in the general model have also become apparent. For example, it is clear in real-life situations that few users of common property are so unencumbered by societal restraints that they can behave as selfishly as the model predicts. Also, contrary to one of the model's important assumptions, when resource levels begin to decline, individual users of common property usually have some awareness of their part in the impact on resources (see Berkes, ed. 1989; McCay and Acheson, eds. 1987; Cordell, ed. 1989).

Another erroneous assumption implied in the tragedy-of-the-commons model is that private property will always be more diligently conserved by its owners, an assumption that is often the linchpin for proponents of privatization in the fisheries. However, as several scholars have pointed out, private property status provides no absolute assurance that resources will be conserved. In countless cattle-raising and agricultural regions around the world, the impoverished private owners of lands have overgrazed their pastures and exhausted their agricultural soils (see Ebenreck 1984; Gilles and Jamtgaard 1981). Such conditions, McEvoy (1986: 171) notes, are often "less a cause of poverty and powerlessness than its symptom."

Still another important defect in the model is ascribing the causes of environmental degradation and economic deterioration "to the nature of property rights instead of acknowledging the role of more complex features of socioeconomic systems" (McCay and Acheson 1987: 9). The depletions of resources that have heretofore been associated with common property rights, may in fact be more a result of colonial or capitalistic policies or of industrialism and modernization. In this vein, R. W. Franke and B. H. Chasin (1980: 120–22) attribute much of the tragedy in the management of common property resources in the Sahel to the loss of a traditional local "bioethic" with the transition to colonial production modes. The new bioethic of colonialism preached "unbridled personal accumulation," rather than self-sufficiency and restraint in the use of common property resources (see also Emmerson 1980).[1]

What I find most objectionable about the tragedy-of-the-commons model, at least when it is applied to the fisheries, is the cynical view of the mentality, character, and personality of fishers implied in the explanation of how the tragedy develops. That view essentially assumes that as overall yields for a given level of effort dwindle, fishers inevitably develop a greedy, "take all you can, and take it now" approach to the fisheries. As McEvoy (1986: 253) appraised this point of view, "The farmers in Garrett Hardin's 'tragedy of the commons' are fundamentally autonomous, self-serving, irresponsible creatures, as radically alienated from each other as they are from the grass on which they feed their cows."

Real-world fishing people, it turns out, often act contrary to what Hardin's theory predicts. In a psycho-cultural study conducted in western Puerto Rico, for example, Poggie (1978) found that small-scale fishers of very modest means scored higher in deferred gratification than their non-fishing peers in the same community. His study also revealed that those fishers who scored highest in deferred gratification were also the most successful ones in their communities. From this, Poggie concluded that a deferred orientation seemed to be of adaptive significance in small-scale fisheries. His findings are congruent with those of an earlier study by Conrad Phillip Kottak (1966), who also observed the greatest restraint exercised by the more successful small-scale fishers in Brazil.

Poggie made an even more important discovery, though, namely, that the deferred-gratification orientation declined among the Puerto Ricans who worked on larger, company-owned fishing boats. These findings were also congruent with the conclusions of an earlier study, that of Richard B.

---

[1] If one of fishery science's great paradigms is so flawed, imagine how many other, similarly axiomatic assumptions about fishers might also be mere echoes of outdated and oversimplified points of view.

Pollnac and M. Robbins (1972). Based on his evidence, it seemed to Pog-gie that deferred gratification increased, leveled off, and then decreased as the degree of modernization increased.

In this regard, it seems that intense and increasing competition—par-ticularly that posed by participation in modern market systems—more than any inherent flaw in the character of fishers themselves compels the behavior associated with the tragedy of the commons. Local fishers will indeed stop fishing with restraint when newcomers invade their fisheries. Similarly, local fishers will feel compelled to abandon local traditions pre-scribing restraint when they become caught up in market systems that put their very survival at risk. But the fault in such cases clearly lies not in the human nature of fishers or in their fisheries being common property, but elsewhere.

It also seems clear that a fishery's status as a common property resource is ultimately not as deleterious as its institution as an open-access com-mons. As I shall discuss in a later chapter, when local fishers own a particu-lar fishery in common and can also restrict access, they often voluntarily institute means for limiting fishing effort and fish mortality among them-selves—even when they supply their catches to a modern market. So from this perspective, "the tragedy of the commons" should perhaps be re-named "the tragedy of open access" (see Berkes et al. 1989: 93).

In this same vein, Bell (1978: 137) makes the following point: "The common property nature of the fishery resource has nothing to do with the behavior of the marginal and average catch per unit of fishing ef-fort. . . . The private or sole owner of a fishery resource would still be faced with these same relations; that is, he would face dwindling produc-tivity of the resource as exploitation increased." Of course, many fisheries experts still feel that privatization is the best way to induce fishers to re-duce their fishing effort. Richard J. Agnello and Lawrence P. Donnelley (1975), in their study of the U.S. oyster industry, observe: "The empirical findings suggest that private property rights do in general make a signifi-cant difference in a state's average labor . . . productivity in oyster harvest-ing. Common property rights are associated with low labor productivity resulting from . . . over exploitation." But these comments pertain to a fishery in the United States where private property is a strong institution. They might be absolutely inapplicable in a different cultural milieu.

As we shall see in later chapters, fishing effort is indeed constrained by cultural practices in various local common property fisheries. Many of these practices were developed in the seesaw, trial-and-error process of cultural adaptation, as well as within the context of the larger culture's history and great traditions, and therefore they often fit nicely into the larger fabric of the national cultures of which they are a part. Moreover, in many instances such local or indigenous practices help to prevent the development of the tragedy of the commons.

A Time for Reappraisal

Because the validity of the long-standing assumptions underlying the tragedy of the commons is increasingly in question, at least so far as the fisheries are concerned, it is important that the model now be seriously reappraised. More fruitful, it seems, would be to shift the emphasis from common property as an institution to the human element—to an examination of what motivates fishers to fish "as if there were no tomorrow." However useful the tragedy-of-the-commons model may have been in helping fisheries managers to conceptualize fisheries problems, it seems also to have reified some erroneous assumptions about fishers that have led to inappropriate fisheries policies.

Institutionalized theories, like institutionalized bureaucracies, are slow to change. Nevertheless, recent studies that mitigate the determinism of the tragedy-of-the-commons model are beginning to have an impact on our thinking about fisheries policies and fisheries management. Undoubtedly some of the most important of these studies are those that demonstrate the reality and ubiquity of effective local systems of self-management, or indigenous management, in various fisheries around the world. "Thus," Svein Jentoft (1985: 327) happily concludes, "the fishery does not always resemble a game like Hobbesian anarchy as presumed when Common Property Theory is applied."

With these recent revelations, fisheries management theory seems to be entering a new era, one in which a consideration of local management practices among small-scale fishing peoples promises to play a larger role in the development of fisheries policies.

A New Era in the Fisheries

For all the fishing nations, a sober-ing reality struck in the late 1960's and early 1970's: the world fish catch, it was generally conceded, was approaching its upper limit and was not likely to increase much more in the near future. By then, the distant-water fleets of most of the principal fishing nations had already begun to suffer disastrous economic reversals with the virtual collapse of several major fish stocks. And by then, too, an increasing number of domestic small-scale fishers found themselves being marginalized by foreign fishers operating just offshore.

A new era in the fisheries had dawned, one that would see the world's coastal nations zealously attempting to reserve for themselves what had for so long been common property, open-access resources. The international politics of the ocean, till now predicated on abundance, became predicated on scarcity.

The Freedom of the Seas

The phrase "the freedom of the seas" rolls easily off the lips of many landlubbers. It asserts that nobody can own the oceans and seas, that their resources are the common property of all of us. Sometimes this is carried further, to embrace a freedom from the many encumbrances of life ashore. But to consider this doctrine of freedom naturally appropriate for the watery realms is a grievous error. Indeed, when people assert the freedom of the seas they are projecting a romantic ideal of freedom from ownership and behavioral restraint that is hardly ever seen in the functioning of human societies on land. To assume the naturalness of the doctrine in the use of the fisheries is equally fallacious, since these are merely regions where human societies have extended themselves beyond the shore.

Actually, the doctrine of the freedom of the seas has a decidedly less romantic and more prosaic origin. It only became a hard doctrine during the Age of Exploration, when European navigators went forth to explore the globe and claim new lands for their monarchs. One of the first instances of its articulation can be seen in a stern letter that Queen Elizabeth I of England wrote to the Spanish ambassador in the year 1580, just eight years before her navy defeated the Spanish Armada. "The use of the sea and air is common to all; neither can a title to the ocean belong to any people or private persons, forasmuch as neither nature nor public use and custom permit any possession thereof" (Bartlett 1968: 188).

These ideas were further advanced in 1604 by a young attorney in the employ of the Dutch East India Company named Hugo Grotius. Grotius had been retained to defend a ship captain working for the company who had seized a Portuguese galleon in the Malacca Strait. The Portuguese argued that the "Eastern Waters," or the East Indies, were their private property; Grotius contended that the seas were free and open to all nations. Grotius won his case, but as one scholar notes, "That judicial victory would have meant little were it not for the ambitions of the rising British and Dutch empires and the power of their fleets" (Pontecorvo, ed. 1986: 30).

A few years later, in 1609, Grotius published and further explicated his ideas. In a pamphlet titled *Mare Liberum*, he declared that the "high seas" were "for the innocent use and mutual benefit of all" (cited in Fye 1977: 2). Elsewhere he wrote that "the sea, since it is as incapable of being seized as the air, cannot be attached to the possessions of any particular nation" (cited in Pontecorvo, ed. 1986: 30).

Thanks to the pioneering efforts of the great European explorers—Columbus, Vasco da Gama, John Cabot, Magellan, Cortez, Drake, and the rest—transoceanic crossings became almost as routine by the nineteenth century as transoceanic jet airplane travel is today. Most of the known world, at least its more favorable regions for human habitation, had been claimed and occupied by Western European peoples, whose new colonial empires bustled with activity.

The conquered and colonized indigenous peoples found all their prior claims to territories and resources superseded, including their marine resources. So what if the Tahitian chiefs claimed ownership of all the seafoods around a certain atoll and prohibited harvesting them during certain times of the year? There were ships' stores to replenish, new settlements to feed, and soldiers with swords, muskets, and cannons aplenty to make sure the work got done. It simply was not in the colonial mentality to consider a different order of things.

The indigenous peoples who survived and adapted to the new colonial regimes were acculturated to the ways and mentalities of the conquerors. Whole cultures thus went extinct as succeeding generations learned less and less about the old lifeways. As a result, a wealth of native knowledge about the managing and conserving of natural resources—including marine resources—was forgotten, passing into extinction along with the cultures that had accumulated it.

Neocolonialism and Internal Colonialism

Western Europe maintained control over its colonies with varying degrees of success for nearly three centuries. Some feel this colonial era had run its course and was brought down by the nationalist movements of the nineteenth and twentieth centuries. Now, however, it is clear that the old colonialism merely metamorphosed into new forms, in which colonial-type relationships persisted between nations as well as within them.

True, those fervent nationalist movements did see most of the colonized nations transformed into sovereign states. But while the new regimes were often successful in regaining control over their terrestrial resources, for those with marine resources, the process proved to be difficult, since most of them lacked sufficient naval power to exclude foreigners from the waters beyond their shores. Moreover, even when they stridently objected to foreign incursions into these realms, the foreigners often merely countered by continuing to assert the doctrine of the freedom of the seas.

The old colonialism, rather than dying out, merely transformed itself into two new forms: neocolonialism and internal colonialism.

In essence, neocolonialism is the political, economic, and cultural domination of the poorer and developing nations by the developed ones. In connection specifically with the fisheries, neocolonialism means that fishers in the developing nations cannot by themselves keep foreign fishers out of their waters or protect their fish markets. Moreover, fishing peoples who live in attractive coastal settings are often powerless to prevent foreign corporations from developing tourist industries in their midst—industries that may not only interfere with their fishing activities, but subvert their most basic traditions and values.

Threatening as all this is to the fishers in these nations, they are often threatened even more by internal colonialism, defined as an institutional-ized social, political, and economic order that enables the rich and power-ful members of a national society—mainly urbanites who control the na-tion's political and economic high ground—to exploit their rural areas (González Casanova 1965). Generally speaking, although nationalist move-ments have helped developing nations to resist or mitigate the effects of foreign domination and neocolonialism, most have fallen short in address-ing the matter of internal colonialism. Overcoming what developmen-talists for good and circumspect reasons are wont to call a country's "inter-nal structural problems" requires a radical restructuring of the nation's social, economic, and political status quo, something that is unlikely to be brought about without civil warfare.

Franke and Chasin (1980) provide a penetrating examination of how first colonialism, and then its perpetuation in colonial-type relationships in the modern era, led to the widespread famine that struck the West African Sahel in the late 1960's and early 1970's. Though the phenomenon they describe pertains mainly to a landlocked area, it nevertheless provides im-portant lessons for any developing nation interested in developing its fish-eries—a problem concerning which I will make some modest suggestions in the final chapter.

During the original colonial era in the Sahel, developments beneficial to the colonial interests—particularly monocropping on a large scale—were destructive to the natural environment and eliminated many of the diverse sources of subsistence that had always sustained the native peoples in the past through the droughts that periodically afflict the region. In addition, a new emphasis on export crops meant that relatively less food was avail-able when the droughts struck.

According to Franke and Chasin (1980: 85–91), colonialism left the Sahel five legacies that increased its dependence on the developed nations and its vulnerability to famine: (1) a lack of economic diversity and an em-phasis on exports, which made the region more vulnerable to increasingly unfavorable terms of trade, in which export prices never rose as fast as the prices of essential imports; (2) dependence on foreign funds for invest-ment, much of which came from multinational corporations; (3) increased dependence on the business cycle—which is mainly driven by the devel-oped nations—and which makes the region more vulnerable to the boom-and-bust cycles of capitalism; (4) dependence on the outcomes of rivalries between the capitalist nations, which impacts the markets for the Sahel's exports; and (5) the persistence of internal colonial relationships under the guise of nationalism.

The Dark Before the Dawn

The culmination of the second era in the fisheries—the era that was launched by the Industrial Revolution and that ended when the world fish catch leveled off—was epitomized by the great fleets of factory ships that roamed the North Atlantic from the 1950's through the 1970's. Their approach to fishing was a radical departure from any the world had seen before.

William W. Warner, in his wonderful book *Distant Water: The Fate of the North Atlantic Fisherman* (1983), provides an intimate description of these fleets, one in which we go on voyages and share the thoughts and activities of the fishers who took part in them—fishers hailing mainly from the United States, Great Britain, West Germany, Spain, and the Soviet Union.

The ships were huge, some exceeding 300 feet in length and 4,000 gross tons. Many were capable of bringing in as much as 500 tons of fish in a single haul and of processing over 250 tons a day. Because of their formidable capitalization requirements—a purchase price of several millions of dollars and daily operating expenses in excess of $20,000—they were usually underwritten by giant corporations and heavily subsidized by national governments.

These fleets specialized in fishing in cold waters, where the largest aggregations of single species in the sea are found, catching mainly fish that were in high demand in European and North American markets. Because of their processing capacity, the ships could stay at sea longer than any fishing vessels had ever been able to before—sometimes for a year or more. Each was a floating factory, and the fleet itself a kind of roving industrial complex, where even the by-catches of untargeted species and the scrap from the on-board processing operations were reduced to fish meal.

In 1974, according to Warner (1983: 53), the Soviet fishing fleet alone—then the largest in the world—contained 710 factory trawlers, 103 factory or mother ships, more than 2,800 smaller side trawlers, and more than 510 support vessels, including refrigerated fish carriers, oceangoing tugs with on-board repair shops, research and scouting ships, supply ships, and fuel tankers. Warner presents a graphic image of the Soviet fleet in its heyday as it swept up and down the Georges Bank fishery off the New England coast. First concentrating on cod and herring, and then, as those stocks declined, turning their attention to haddock, the Russian ships "paced out in long diagonal lines, plowing the best fishing grounds like disk harrows in a field" (p. 6).

Once the North Atlantic fleets were equipped with new sensitive and sophisticated fish-locating electronics, as well as such technologically advanced gear as midwater trawls, they no longer had to target merely ground fish and could go after fish anywhere in the water column. Moreover, thanks to an improved capacity for reconnaissance and communication, when one vessel found a particularly large concentration of fish—a spawning school, for instance—practically the entire fleet could quickly converge on it and work it until it was all but wiped out.

It was this practice, which came to be known as *pulse fishing*, as much as anything else, that led to the catastrophic depletions of the main fish stocks in the North Atlantic. Moreover, pulse fishing worked severe hardships on the near-shore fishers who had traditionally exploited migratory stocks when they moved into near-shore areas (Warner 1983: 57). For these reasons, pulse fishing has now been curtailed or severely restricted in most ocean regions.

The North Atlantic fleets reached their peak in 1974, when "1,076 Western European and Communist-bloc fishing vessels swarmed across the Atlantic to fish North American waters" and caught 2,176,800 tons, "ten times the New England and triple the Canadian Atlantic catch" (Warner 1983: 58). Ominously, however, the catch per unit of effort did not reach the level of 1968, when a slightly smaller fleet had harvested around 2,400,000 tons. Moreover, the fish that were caught had been getting smaller and smaller since 1968, "nearly always an early and sure warning of the general decay of major fishing areas" (Warner 1983: 58–59). The North Atlantic fishery had signaled that it had no additional fish to give.

In the meantime, fleets of factory ships roved other waters—the Japanese in the North Pacific, for instance. Much like their cousins in the North Atlantic, they concentrated their efforts in cold waters, where large aggregations of single species could be taken, and fished fairly close to shore, where the greatest concentrations of fish were found. By the early 1970's, they too were experiencing difficulties in increasing their overall catches.

For many of the small-scale fishers who lived along the coastlines that were being exploited by these roving fleets, the result was economic disaster. Not only did the offshore fleets deplete the stocks they had traditionally depended on, but the industrial fishers could usually outcompete them because of the economies of scale they enjoyed.

Something had to give and finally did, with the break underscored by two events. First, as we have seen, the oceans and seas signaled that they could sustain no further production increases, at least not in the northern cold-water fisheries. And second, the world's coastal states began to assert their exclusive rights to extended zones far from their shores in an attempt to drive foreign fishers out of these regions.

As Warner (1983: ix) notes, these events would "undermine the economic viability of the distant water fleets and their costly factory trawlers." Because they were "unable to adapt to the changing environment of world fishing they themselves did so much to create" (p. 309), their numbers quickly declined. Most of the largest factory ships were refitted for other purposes or scrapped. The few that managed to stay in business turned their efforts to warmer waters and were refitted to harvest the greater variety of species found in those regions.

So factory ships still rove the seas and still collectively take big catches. And new problems have arisen in conjunction with the development of more extensive fishing operations. Drift-net fishing, for instance, particularly in the Pacific, now sees vessels from roving fleets deploy huge gill nets, up to 30 miles long, which indiscriminately ensnare most of the marine life they intercept.

The United Nations Law of the Sea

Recent and radical changes in legal regimes for managing ocean resources have shifted increasing attention to small- and intermediate-scale fishers. Chief among these new regimes is the United Nations Law of the Sea (UNLOS-82), which was released for signature in 1982. Though this convention was still awaiting ratification in 1989 and was technically not yet in force, it has been accepted by most nations as the new international law of the sea, especially its provision that grants the world's coastal nations exclusive rights over living marine resources within 200 miles of their shores. This new 200-mile limit, which is called the *exclusive economic zone*, or EEZ, marks the end of the doctrine of the freedom of the seas, which heretofore had allowed anyone with the wherewithal to operate in most of the world's fisheries for nearly four centuries—at least assuming they complied with the provisions of various multinational treaty agreements established for certain fisheries (UN 1982; Pontecorvo, ed. 1986).

By common agreement, coastal nations could formerly only assert exclusive rights to ocean resources fairly close to their shores—typically no farther than twelve miles out. But UNLOS-82 makes it illegal under international law for any fishers to extract living resources from another nation's EEZ without first getting permission and making certain payments and other concessions. The ocean waters beyond the 200-mile limits essentially remain the common property of all the nations of the world, although even there the freedom to fish is still constrained through various multinational treaty agreements. As Warner (1983: 320)) has commented, "The richest meadows of the sea—the continental shelf and slope waters that are home to eighty-five percent of the world's harvestable fish—are now a staked plain."

The promulgation of UNLOS-82 was merely another step in a worldwide movement to change fisheries tenure that began in the early days of the twentieth century and intensified considerably following the Second World War. The movement was marked by a variety of unilateral declarations on the part of several coastal states asserting their exclusive rights to marine resources close to their shores, as well as by the formation of multilateral fisheries agreements and the proliferation of multilateral commissions with such formidable acronyms as ICNAF, IATTC, and IWC, organized mainly to regulate various high-seas fisheries.[1]

The first postwar move in what amounts to the biggest "sea grab" since the Age of Exploration was made by the United States. The "Truman Proclamation" of 1945 was aimed primarily at protecting the oil and gas reserves situated on the country's continental shelves. Nevertheless, the proclamation did more than claim exclusive jurisdiction over hydrocarbons. It extended the claim to living resources situated over the continental shelves and even to some territories beyond them that were contiguous to coastal areas. This proclamation, McEvoy (1986: 192) notes, "ran counter to a longstanding U.S. tradition of support for free access to international waters."

Several other countries soon followed suit, unilaterally claiming exclusive jurisdiction over marine resources within 200 miles of their shores. Chile took the step in 1947, followed quickly by Peru the same year; Ecuador staked its claim in the early 1950's. These countries, which all have narrow continental shelves, were particularly concerned with protecting the rich fisheries lying just off their coasts. After Korea claimed jurisdiction to 250 nautical miles in 1952, several other coastal states hurried to make their own declarations.

Ironically, for more than two decades after its move, the United States

[1] Respectively, the International Commission on Northwest Atlantic Fisheries, the Inter-American Tropical Tuna Fishing Commission, and the International Whaling Commission.

staunchly opposed these claims by other nations on the grounds that the extended jurisdiction would constrain its naval power, hinder its shipping, and reduce the productivity of its distant-water fishing fleets, particularly those harvesting tuna and shrimp. But finally, in 1976, the United States reluctantly joined the others, asserting jurisdiction over its fisheries out to 200 miles.[2]

Coinciding with this postwar wave of unilateral moves was a United Nations–sponsored series of international meetings aimed at drafting new regulations on the use of the seas and their resources. At the first of these law-of-the-sea conferences, held in Geneva in 1958, no consensus was reached on what the new territorial limits should be. Some nations argued that the limit should remain at three miles, and others that it should be extended to six or twelve miles. Several wanted to see it extended considerably farther. The matter was not resolved by successive conferences in 1960, 1967, 1974, 1975, 1976, and 1979, but the participating states did manage to flesh out the meaning of certain instrumental concepts, such as rights of ownership, rights of access, territorial seas, contiguous zones, continental shelves, high seas, and living resources.

As we have seen, the new United Nations Law of the Sea treaty was finally released for signature in 1982. It was born into controversy, however, mainly because of unresolved issues concerning rights to exploit minerals on the deep sea floor. And it still does not have the force of international law, though practically all member nations have agreed in principle with the provisos that establish the 200-mile EEZs.

UNLOS-82 is a particularly important milestone for the formerly colonized and now developing nations. Indeed, worldwide recognition of the EEZs has already benefited many of the fishers who had found themselves in a losing competitive struggle with foreign fishers. Moreover, with the EEZs secure, the developing nations are now in a good position to plan and implement their developmental aspirations for their fisheries. Unfortunately, UNLOS-82 will do little to help them address the problem of internal colonialism.

In all the world's coastal nations, rich or poor, worldwide recognition of the EEZs has created immense new challenges for fisheries management.

---

[2] The 1976 act was originally called the Fisheries Conservation and Management Act of 1976, or simply FCMA-76, and was codified as Public Law 94-265, 90 Statute 341-361, U.S. Congress. It was renamed in honor of its principal architect, Senator Warren Magnuson, in 1981 and is now referred to as either the Magnuson Act or the MFCMA. The act has been amended twice, in 1981 and 1983.

By laying claim to these 200-mile zones, the world's coastal states have collectively acquired over 25 million square miles of ocean territory that they must now administer for the general welfare of their citizens.

The acquisition of exclusive property rights over marine resources within what amounts to around a third of the total ocean surface (Fye 1977: 4, 19) has naturally led each nation to focus more consciously on its small- and intermediate-scale fishers, nearly all of whom are found within the EEZ. Thus UNLOS-82 has by implication prompted a greater concern for small-scale fishers than has ever been seen before.

This increased attention has brought a new appreciation of how important these groups are to many national economies. Right now, for instance, over half the world's fish catch is obtained in coastal waters. Moreover, because small- and intermediate-scale fishing tends to be devoted to producing human foods and to be more energy efficient than large-scale fishing, its potential cannot be minimized, given current projections of population growth and energy shortages.

In general, unlike the energy economics of distant-water fishing, the energy economics of producing high-quality animal protein in coastal waters—expressed in terms of the production to energy expenditure ratio—compares very favorably with that for pastured cattle. Moreover, as we have seen, it is mainly large-scale, industrialized fishing that produces the 30 percent of the world catch that is converted into poultry and cattle feed (Idyll 1978: 10–12, 128). On the other hand, one comparative disadvantage of small-scale fishing cannot be denied: its fishers lose a greater proportion of their catches to spoilage.

One unfortunate and unforeseen result of extending jurisdiction over fisheries out to 200 miles is the emergence of internal conflicts within many countries—conflicts between sectors of fishers, conflicts between fishers of all types, fisheries administrators, scientists, and so forth. In the United States, for example, the New England fishers strongly favored the 200-mile limit and the anticipated elimination of foreign competition. Most assumed that extended jurisdiction would enable them to increase their fishing effort. But scientists and administrators saw the new limit as an opportunity for retrenchment and the application of more stringent conservationist measures. M. Estellie Smith (1982: 90) quotes a disillusioned New England fisherman who undoubtedly spoke for most of his peers: "Face it," he said, "as soon as that 200 mile limit went into effect the Feds owned the fish."

A New Era for Small- and
Intermediate-Scale Fishers

The new law of the sea is already prompting the modernization of fisheries policies worldwide by compelling the various coastal states to develop new policies for their EEZS. Moreover, the UN's Food and Agriculture Organization (FAO) is devoting considerable effort to improving the living standards of small-scale fishing peoples, raising their productivity, and improving their means of fisheries management (UN 1982).

This new impetus toward smaller-scale approaches to fishing, perhaps more than anything else, is what most heralds the beginning of a new era in fisheries management. Gone is the primary concern with fish and large-scale fishing enterprises that characterized fisheries management in its earliest days. Increasingly the primary concern has become fishing peoples.

While the primary goals of most fisheries managers today are still to maximize the biological productivity (MSY) and economic yield (MEY) of their domains, a third and more elusive goal has fairly recently been proposed: management for *optimum social yield*, or OSY, with "optimum" meaning the optimum sustainable well-being for a fishery's human participants. In theory, OSY incorporates both MSY and MEY, and the concept is sometimes even extended to cover the larger society as well (see Emmerson 1980; Roedel, ed. 1975; Schaefer 1975).

Admirable as the goal of OSY is, it poses a so-far insuperable problem for fisheries managers: hardly anyone can agree on what the optimization of human well-being entails, much less how to attain it. Consequently, most managers still rely on MSY and MEY as the prime objectives of their policies, even though, as we have seen, those concepts too are problematic.

In the United States, formal recognition of the need to take account of the human element in the management of fisheries came with the passage of the Magnuson Act, which used this very term, optimum sustainable yield, in mandating the managers' work. Even though OSY in this case was conceived as the optimization of biological as well as human social and economic objectives, it is fair to say, as Maiolo and Orbach (1982: 3) do, that "social analysis is mandated . . . in the U.S. Fishery Conservation and Management Act of 1976." Unfortunately, the fisheries management establishment in the United States has been slow to act on this mandate, mainly because there is still so much confusion about how to define and operationalize OSY.

Humanizing Fisheries Management

As we have seen, when modern fisheries management was first developed, an important consideration was overlooked: how local peoples often regulated their fishing effort on their own. Can indigenous techniques not tell us something about the way we might achieve the elusive "optimum social yield"?

Now discovering indigenous means of management is no easy task, since few fisheries today are totally free of external regulatory regimes. Moreover, any lesson that we might learn from local means of control, particularly those that we can discover only by examining the historical record, may be valueless for managing those fisheries where producers are already locked in a deadly competitive struggle to supply a high-demand market with an increasingly scarce resource. Locally developed management regimes often break down and will not work when a fishery becomes linked to an outside, modern market system. Nevertheless, the task is well worth the effort, not only because where indigenous management means are still in force their demise would be a grievous loss for the societies that have developed and employ them, but also because they provide important clues to how more humanized regimes for fisheries management might be developed in the future.

In defining indigenous means of management, I take a broad, all-inclusive approach. I mean any people's practice, whether conscious or unconscious, intentional or inadvertent, active or passive, recognized as resource regulation or not, that has the effect of limiting the mortality of marine resources resulting from fishing effort.

Indigenous management is mainly the purview of small-scale fishers, and in the literature pertaining to them it goes by various names. Some of the more commonly encountered synonyms or closely related concepts are "community-based management," "localized management," "organic management," "self-management," "bottom-up management," "traditional management," and "traditional sea tenure." For many researchers, particularly maritime anthropologists, the term is practically synonymous with local systems of "sea tenure," defined by Ruddle and Akimichi (1984: 1) as "the ways in which fishermen perceive, define, delimit, 'own,' and defend their rights to inshore fishing grounds."

As McCay (1981a: 5–6) notes, "Most known cases of indigenous fisheries management hinge upon the management of access to fishing *space* rather than levels of fishing effort." She singles out three basic means or strategies of indigenous management. One is the assertion of formal or informal property rights over prime fishing spaces. Another is the devising of ways to exclude newcomers or outsiders from a territory deemed to be one's own. And the third is what can be called an information-management strategy, which in effect confers temporary property rights to certain fish stocks. McCay further notes the difficulties local fishers face whenever they try to limit fishing effort on migratory stocks that wander out of their spheres of control. And like Johannes (1977), she is concerned about "the vulnerability of indigenous fisheries management systems to attrition or destruction by forces of economic and technological change and by the power of the state and the international systems."

Indigenous management, as I define it, must be understood as part of a people's fisheries technology, technology being used in a broad sense to include not only the tools and other hardware people employ to exploit their fisheries—but also their knowledge about how to put that hardware to best use, how to organize fishing activities, and so forth. Technology so conceived therefore consists of many components of ideational culture that are generally understood among members of a particular society. Any means of indigenous management that a people employ—whether consciously or unconsciously—are thus embedded in their cultural and economic systems.

Many fishing peoples still strive to regulate their fisheries, often even when externally imposed management regimes have officially superseded their indigenous ones, but even now, after almost a century of modern fisheries management, we do not know very much about these practices. Fortunately, the body of literature on this neglected topic has been growing rapidly in recent years, so rapidly in fact that Ruddle and Akimichi (1984: 1) happily state that fishers' systems of sea tenure are "one of the most significant 'discoveries' to emerge from the last ten years of research in maritime anthropology," even if "they are nothing new to fishermen." [3]

So far this body of literature has not made much impact on the modern practice of fisheries management. One problem is that most of the works

[3] For important pioneering works on local indigenous management systems, sea tenure, and so forth, see Johannes (1977, 1978) and Klee, ed. (1980: 245–81) on Oceania; and Andersen and Stiles (1973) and Andersen (1979) on Canada. The new interest in the subject has given rise to such short studies as Berkes (1985); Dahl (1988); Davis (1984); Durrenberger and Pálsson (1987a); Jentoft (n.d.); and McGoodwin (1984). It has also led to several book-length works, including most notably Berkes, ed. (1989); Cordell, ed. (1989); McCay and Acheson, eds. (1987); Ruddle and Akimichi, eds. (1984); and Ruddle and Johannes, eds. (1985). An excellent summary and critique of Ruddle and Akimichi, eds. (1984) appears in Durrenberger and Pálsson (1988).

are still so new that they are not widely known. But another, more serious problem is adding to our fund of knowledge: few of these indigenous systems are formalized or written down; many are merely implied through historical practices; and a good number are covert or illegal. Moreover, as noted earlier, most of the existing studies use a diverse and nonuniform array of methodological approaches. The bulk are mere existence demonstrations, showing that a wealth of indigenous management practices, both historically and at present, have been in effect. Few have sufficient quantitative salience to permit an assessment of the relative effectiveness of the systems discussed. So we cannot really blame modern fisheries managers for a lack of interest in what they might otherwise be favorably inclined to consider as viable alternatives to prevailing practices of management. Let me emphasize once again, before I go any further, that there is no reason to suppose that such "bottom-up" forms of fisheries management are inherently superior and preferable to systems imposed from the top. Especially in the modern era, self-interest and short-term considerations often override the best intentions among fishers to manage things for the good of all. Moreover, indigenous means work best mainly among homogeneous groups of fishers who employ similar types of fishing gear, but hardly at all where there is significant stratification, as there is in many fisheries today (Bailey 1988b: 119).

The point is, until we learn about these techniques, we cannot tell what lessons they may hold for future fisheries policies. In the next two chapters, I will discuss passive and active means of indigenous fisheries regulation. *Passive means* are those cultural practices that may have been instituted without any intent to regulate the fishery, but that nevertheless have the effect of limiting fishing effort or fish mortality.

Now, some readers may object to classifying passive means as instances of self-regulation, arguing that what we should focus on are the (active) means that fishing people consciously employ for the specific purpose of regulating their fisheries. But passive means are important whenever they limit fishing effort and fish mortality. As I mentioned in an earlier chapter, many beneficial cultural practices have become customary in fishing societies and have persisted even though the original rationale for instituting them has been lost from memory.

My classification of various indigenous fisheries management strategies as passive or active is somewhat arbitrary and does not correspond with any clear distinctions in reality. A means of passive regulation, especially when it is extremely effective, might in fact merit classification as an active strategy instead. Thus from a conceptual point of view, the division of the material that follows into two chapters should not be taken as particularly important. Again, what *is* important is to appreciate that if these indigenous practices help to reduce either fishing effort or fish mortality, they may be worth incorporating into modern fisheries management regimes.

Passive Means of Indigenous Regulation

M any local fishing societies successfully control important marine
resources through ritualized or habituated patterns of behavior that were
shaped in such small increments and over such long periods of time that
their members are only barely conscious of them. However, if these patterned behaviors can be reasonably construed to play a positive role in conserving marine resources, and if in their ethnographic settings they seem
to be generally agreeable to the members of the local society who employ
them, then fisheries managers should seriously consider incorporating
them into their management policies.

In what follows, I describe some indigenous means of passively limiting
fishing effort or fishing mortality that may offer useful insights for modern
fisheries management. Unfortunately, few of the studies in which these
means are discussed contain quantitative data that would allow a reasonable estimate of the degree to which they reduce resource mortality.

Inability

High on the list of passive means that help to prevent overharvesting or
severe competition is simple inability: a human population too small to do
much damage because it makes little demand on marine resources owing
to a low-impact capture technology or to a low-demand market system, or
to some other factor or combination of factors that limits a people's ability
to overpressure a fishery.

This signals an important consideration in fisheries development policies: that when a fishery cannot be left alone, it should at least be developed very slowly and carefully. Whatever course of development is to be

promoted, its planners should keep in mind the factors that have worked in the past, attempting to retain them or their analogs in future management schemes.

Much has been written about the need for development to be "appropriate" to the cultural milieu in which it is introduced. E. F. Schumacher (1973) is a landmark work in this regard, and his thesis that "small is beautiful" is supported by countless case studies describing well-intentioned development projects that brought in inappropriate innovations, much to the target population's detriment.[1] Also, programs or policies that strive to increase the population in coastal zones should be seriously questioned. Currently in many developing nations less-populated coastal regions are targeted for resettlement by impoverished agriculturalists from the interior.

We have already noted a case where just the opposite occurred. When the aboriginal population along North America's Pacific coast was decimated by the incursion of new settlers in the late eighteenth and early nineteenth centuries, their removal caused, in the words of Gordon W. Hewes (1973: 149), a "non-deliberate conservational effect" that allowed the region's salmon stocks to recover. As a result, when the commercial salmon industry began to develop in the 1860's, the stocks were higher than they had been for millennia.

It would be unthinkable, obviously, to propose the deliberate decimation of a human population or even to stand idly by and let war, famine, or epidemics bring about a reduction of fishing effort in hard-pressed fisheries. Nevertheless, the importance of reducing or limiting the population in coastal fisheries by other, more humane means cannot be overemphasized.

Some people feel that societal attempts to limit population growth are immoral, an infringement of basic human rights, and most people regard state-sponsored attempts to limit population as threatening and offensive. The general sentiment that population control should ultimately be a matter of personal choice was given voice in 1967, when the United Nations asserted: "The Universal Declaration of Human Rights describes the family as the natural and fundamental unit of society. It follows that any choice and decision with regard to the size of the family must irrevocably rest with the family itself, and cannot be made by anybody else" (U Thant 1968). But the fact is, the societal restriction of population growth has many precedents. Aboriginal societies in particular often took care to see

[1] Readers interested in the "appropriate technology" and "soft-tech" movements will also find interesting Baldwin and Brand (1978).

that their numbers would not exceed the carrying capacities of the environments on which they depended. McEvoy (1986: 29–30), for instance, describes how various aboriginal Indian societies in California employed a wide variety of means to check their growth, including contraception, abortion, taboos on intercourse during certain times, rewards for celibates, infanticide, and geronticide.

The inability to overexploit resources for technological reasons is another important means of passive fisheries regulation. Having boats or other craft too small to venture very far offshore or unable to carry large nets may keep fishing effort and fish mortality at acceptable levels. Riva Berleant-Schiller (1982: 134) notes of the small-scale fishers in Bermuda, who lack many of the amenities associated with modernization, that "so far, an interaction of environment, artisanal technology, and economy has protected the fishery from depletion and preserved it for local use." Similarly, in warmer climates a lack of ice will limit the duration of fishing trips and may be an important passive means of limiting fishing effort (Cordell 1978).

Hence features that at first make a fishery appear backward and underdeveloped may actually serve it well enough to be retained in a modern regime for managing that fishery.

The use of a diversity of gear may also contribute to an inability to overharvest a fishery. As Cordell (1978) notes, the different types of gear employed by small-scale fishers exploiting an inshore fishery in Brazil naturally causes them to space out their efforts, since each type is used only in the microenvironment suitable to it.

Another interesting case of a related nature is described by Francis P. Bowles (1973), who notes that the lack of harbor space in certain Maine lobster fisheries effectively limits the number of fishers who can enter the fishery, thereby controlling both overall effort and competition.

The inability to exploit a fishery because of environmental constraints or bad weather may also be an important passive means of regulation. Environmental constraints may include reefs, shallow water, severe weather, seasonally high tides, or some combination of these factors. Reef complexes, for example, especially those just under the surface that are subject to strong wave action, may serve as resource-conservation zones, providing refuge for fish stocks and a source of resupply for nearby areas that can be fished. And bad weather conditions may discourage fishers from venturing into certain fishing grounds, thereby conserving the stocks there. We know, for instance, that in prehistoric times the periodic icing over of the sea and rivers in northern Norway forced maritime hunters and gatherers to turn from fishing to hunting (Gjessing 1973). Similarly, the icing of rivers and estuarial inlets forced seasonal cessations of fishing activities among the aboriginal inhabitants of the Pacific salmon-fishing regions along the northwest coast of North America (Hewes 1973). The same phenomenon would of course be operative as a potentially important means of conservation in many fisheries today. Where nature itself acts as a constraint on fishing effort, then, fisheries managers might be wise in certain situations to discourage developments that seek to overcome the problem.

One final and cautionary note on the putative inability of many simple, premodern societies to deplete their resources is in order. McEvoy (1986: 39–40) notes that many of the early settlers in California who helped set policy on the exploitation of natural resources "proceeded under the interrelated assumptions [that] the Indians lacked the technical sophistication either to deplete or to manage their resources, and that the destruction of once-abundant 'virgin' resources was a necessary complement to 'civilized' habitation." But in fact, as he goes on to say, many of these supposedly pristine environments had already been considerably altered by the Indians, and closer examination of their cultures reveals that, though many of them indeed once had the technical ability to overharvest their resources, they had developed active means of self-regulation instead. A similar conclusion can be drawn from Fikret Berkes's (1977) study of the contemporary Cree Indians in northern Canada.

Lower-Bound Limits

Another potentially important passive means of self-regulation is a lower-bound limit on fishing effort itself, the point where declining productivity or profitability causes fishers to turn to other options and leave the fishery alone. In the parlance of modern fisheries management theory, this is called the point of bioeconomic equilibrium. The idea is that when additional fishing effort will only result in economic loss, fishers will presumably be deterred from undertaking it, and some will begin to leave the fishery.

Apart from the obviously harsh implications of relying on this strategy as a passive means of limiting fishing effort, it is problematic because the leave-taking behavior it predicts often does not occur. Many fishers persist in fishing even in the face of consistent and long-term economic losses. Moreover, this theoretical lower-bound limit is rarely the same for all the fishers in a fishery. Some may see and take advantage of other options; some may not. And many other factors, such as the type of fishing gear used, may result in a different lower-bound limit for each fisher. We will return to this point later.

These problems do not keep the concept from being generally valid, however, for we can expect all fishers to perceive a point at which their efforts become so unrewarding that they will have to give up fishing and leave the fishery. That this was the case in prehistoric times, for instance, can be inferred from the fact that once the ancient inhabitants of the Peruvian coast had overharvested the marine resources on which they had depended for subsistence, they turned away from their fisheries and redirected their efforts toward developing agriculture (Moseley 1975). Though they undoubtedly still had some marine resources left, they probably felt it would be futile to continue to harvest them as they had before.

The lower-bound limit has been given different names by different researchers, depending on their area of concern. James A. Crutchfield and Giulio Pontecorvo (1969), for example, speak of "competitive withdrawal," a process in which traditionally cooperative fishers become so ruthlessly competitive as their fishery becomes overfished that some eventually drop out altogether. And McCay (1981a: 4) describes how, as a fishery changes from a basically private property, subsistence orientation to a common property, market orientation, an influx of fishers and the introduction of more effective gear will eventually result in "economic overfishing." That point, she says, will usually be reached before the point of "biological overfishing." So though permitting a fishery to be exploited until it reaches its bioeconomic equilibrium would be an effective means of passively limiting further fishing effort, it would be wiser to curtail the effort before that point is reached.

Economic and Occupational Pluralism

After simple inability and the lower-bound limit, probably the most important passive deterrent to fishing effort is economic pluralism. Here we find fishers turning away from full-time fishing because of other demands or opportunities posed by their cultural systems—other economic activities, ritual and community-service demands, and so forth—anything that prompts a turning away from the fishery when continued fishing might otherwise be rewarding. As McEvoy (1986: 28) notes, this practice spreads "productive effort over a wide range of resources, each in its season." To some extent, the practice parallels the modern fisheries management strategy of closed seasons, in the sense that both give important marine stocks time to replenish themselves.

Economic pluralism marks fishing communities all over the world. Orvar Löfgren (1979, 1982) describes a characteristic case: the fisher-crofters of the North Atlantic fringe, who prior to the modern era left fishing at certain times of the year to take up such activities as agriculture, sheep shearing, forestry, and trading. Economic pluralism is similarly reported in the rural fishing communities of contemporary Newfoundland (Leap 1977; Martin 1979); in Ecuador, where rural people are part-time fishers and part-time farmers (Middleton 1977); and in the Caribbean countries, where small-scale fishers seasonally turn to cutting, burning, and cultivating their swiddens (Berleant-Schiller 1982). In Spain small-scale fishers turn their attentions to the local tourist industry in the off-season (Pi-Sunyer 1977).

Relatively few small-scale fishers rely exclusively on fishing for their livelihood. Not only does having economic means other than fishing increase their security, it also effectively reduces their fishing effort. Recall in the discussion of unregulated fisheries how, in turning to agriculture, the aboriginal Peruvians reduced the risk of famine should their fisheries fail.

From this perspective we see how unwise are the management policies that permit only full-time fishers access to the fisheries. Unfortunately, such provisos are increasingly seen in fisheries management regimes around the world. In their desire to limit or reduce overall fishing effort, policymakers have set these rules without considering the broader socioeconomic contexts in which fishing peoples live, with sometimes disastrous consequences for the local fishing peoples.

Again, Löfgren (1982) provides a prime example. He describes how some of the societies in the North Atlantic regions of Sweden, Ireland, and Scotland were transformed from a stable, landed peasantry of fisher-crofters at the beginning of the eighteenth century to landless tenants whose fishing activities are almost wholly confined to wage-labor work aboard trawlers in the twentieth century. His is a particularly penetrating examination of the economic impact of industrialized fishing on local fishing peoples. Not only has it left them much worse off, but the fisheries themselves are now in dire straits, being both overcapitalized and depleted of resources. Lamentably, one of the favorite management responses to the deteriorating situation has been increasingly to shut off access to part-time fishers.

Making full-time fishing an eligibility requirement for entry or access to a fishery may ultimately bring about or hasten the fragmentation and demise of an otherwise well-integrated plural economic system. Often, as A. K. Craig (1966) observes, this policy arises from a desire to develop an export trade and generate foreign exchange, with no regard for the impact on local traditional communities. Several good case studies support his general hypothesis that the result can be pure disaster for small-scale fishers. Bernard Nietschmann (1974), for one, has described the terrible consequences that followed from the conversion of a Caribbean turtle fishery to a cash-producing enterprise. And in my own work (McGoodwin 1980, 1987), I have described the near-ruination of Pacific Mexico's inshore shrimp fisheries when the emphasis was shifted from fishing for subsistence to fishing for export.

Retaining the diverse sources of livelihood that are the substance of a pluralistic economic system makes good sense in fishing communities, in view of the sporadic nature of production and other risks associated with most types of fishing activity. Surely one of the main reasons the human species has been so successful has been its ability to exploit a variety of ecological niches, avoiding overspecialization. Indeed, Paul Jorion (1988: 152–53) goes so far as to argue that it is a universal "sociological law" that "*no one ever becomes a full-time maritime fisherman other than under duress. . . .* It is not the continuous dangerous nature of the occupation which makes full-time fishing so unattractive, it is too risky in economic terms" (emphasis in the original).

Cultural Traditions, Customs, and Beliefs

Certain customs, religious practices, superstitions, or taboos can also be considered a sort of passive means of restraint. In much of Europe, for instance, fishing was long forbidden on the Sabbath. Eliminating 52 days a year from the fishing calendar could conceivably have brought about a 14 percent reduction in both overall fishing effort and fish mortality. This, interestingly enough, is practically the only means of indigenous regulation that seems susceptible to easy quantification. But the matter is hardly that simple, for the requirement that Catholics abstain from eating meat on Fridays, for example, probably increased the demand for seafood enough to offset the reduction in fish mortality effected by the observance of the Sabbath (Bell 1978: 43).

Both Johannes (1978) and Klee, ed. (1980) report similar types of abstentions from fishing activities in widespread regions of Oceania, motivated by religious beliefs, superstitions, and food taboos during certain times or for certain categories of people. These undoubtedly also reduced overall fishing effort and fish mortality in those regions.

A particularly interesting means of passively limiting fishing effort is any sort of cultural deterrent to the use of marine life as human food. F. J. Simoons et al. (1979) report the existence of such deterrents in widespread parts of the world: among the aboriginal Tasmanian peoples living on the islands off southern Australia, who would not eat scaled fish; in vast areas of northeast, east, and south Africa; in many parts of India and Sri Lanka, where the authors estimate that as much as a third of the population never consumes marine foods; and among landlocked indigenous peoples in the North American Southwest, whose religious beliefs and superstitions compelled them to avoid marine foods altogether. Klee, ed. (1980) similarly reports the avoidance of marine food in parts of Oceania, particularly in the Society and Hawaiian Islands.

In this same vein, Bell (1978: 43–44) notes the taboo against consuming nonscaled fish of all kinds, including shellfish, in the Hamito-Semitic cultures of North Africa. "This prohibition," he says, "is associated with the Judeo-Islamic tradition and usually extends to all areas where Islam is found, whether these include specifically Arab cultures or not." Furthermore, he notes, in many parts of Africa there are special cultural prescriptions against women and children eating fish so as to reserve it for male adults. The potential demand for fish may also be reduced in India and parts of south Asia by the Hindu and Buddhist stress on vegetarianism.

Such cultural practices may be motivated by much the same kind of considerations that prompt the seemingly irrational taboo against killing cattle in India. According to Marvin Harris (1966), such a proscription in that food-deficient country not only preserves cattle for use as draft animals and ensures a supply of the dung that is an important fertilizer in agriculture, but also maintains an emergency source of food should other sources fail. In the same way, a taboo against or avoidance of seafood may help to conserve a reserve food supply that can be drawn on in desperate times, when sheer hunger could be expected to overwhelm the prevailing normative prescriptions against its consumption.

Several scholars have also wondered whether local customs mandating "respect" for living food resources might not also provide a basis for their conservation (e.g., Bishop 1981; Brightman 1987: 121; C. Martin 1978). Some traditional peoples, for instance, prohibit sport or recreational fishing because they regard it as trifling with marine organisms.

Information Management and
Skill Differences

If information management and the perpetuation of skill differences among fishers are active strategies for limiting fishing effort, then not possessing such information or skills amounts to a passive means of limiting fishing mortality. The Brazilian raft fishers described by Shepard Forman (1967), for instance, who are kept ignorant of particularly productive spots by fellow fishers expressly bent on reducing competition at those spots, could be said to passively reduce the fishing effort there. In the next chapter, I will more fully consider information management and the maintenance of skill differentials as actively employed strategies for limiting fishing effort.

Some Lessons for Modern
Fisheries Management

Passive means of fisheries regulation may be important for reducing fishing effort and the mortality of important marine resources. Merely because a means is unintentional, it should not be seen as insignificant by fisheries managers.

Some of the passive means discussed are clearly not suitable ways to put a limit on fishing effort. Allowing a fishery to reach its bioeconomic equilibrium point, for example, is to invite long-term and continuing resource depletion, overcapitalization, stagnation, and other forms of social and economic malaise. Institutionalizing local strategies that attempt to limit information or perpetuate skill differences, though certainly effective in keeping the nonknowledgeable from making the same impact on marine resources as their more capable peers, would be inhumane, affronting our notions of social and economic equity by helping to perpetuate inequality, marginality, and inefficiency.

But three passive means seem particularly effective and therefore worth incorporating into modern fisheries management policies: simple inability because of a low population level, rudimentary gear, and economic pluralism. Also worthy of consideration are those long-standing customs or beliefs that mandate the cessation of fishing during certain times or reduce the demand for marine foods.

A coastal populace, especially in developing nations, is usually the first tier of demand for a fishery's products and the most direct determinant of local levels of fishing effort and resource mortality. Thus any policy prescribing a growth of that population should be supported by impact studies showing that the fishery will be able to sustain the concomitant increase in fishing effort. Moreover, in devising their overall management policies, developing countries would do well to put the stress on keeping the coastal populations low. This may require rethinking and rescinding policies that target sparsely populated coastal zones for resettlement by impoverished peoples from the interior, forcing "marine fisheries, because they are an open access resource, [to] act as a giant safety valve that absorbs surplus labor from other sectors of the economy" (Bailey 1988b: 116).

Developmentalists should resist the tendency to reduce the diversity of gear used in a fishery, since a diversity of gear naturally diffuses the fishing effort. Again, this may require a change of thinking on the part of national planners and policymakers, many of whom seem seduced by new technologies as symbolic of progress against "backward" and "primitive" traditional fishing methods. Such views, often springing from idealistic or unrealistic perceptions of life in the modernized countries, are encouraged by the promotional efforts of manufacturers and sales representatives of fishing equipment.

The strategy of subsidizing infrastructure projects rather than subsidizing fishers directly should also be considered with caution. Enlarging harbor space or developing shore facilities that speed up or increase a community's ability to market its catches, for example, may prompt levels of fishing effort that the local fishery cannot sustain.

The retention of economic pluralism among fishing peoples is usually a very harmonious passive means for controlling fishing effort. Though fisheries managers tend to view economic pluralism as symptomatic of economic marginality (Maiolo and Orbach, eds. 1982: 143), to most fishers it is just the opposite—a means of reducing risks and ensuring a more secure, prosperous, and interesting way of life. Real hardship can and has been worked on fishing peoples forced into full-time fishing as a condition for entry to a fishery. In the not-infrequent cases where the fisheries later suffered a reversal, these fishers were left with no other means of livelihood. Economic pluralism is therefore almost always a healthy state of affairs in the fisheries. Its main benefits are that it effects an overall per-capita reduction in fishing effort while affording fishers more security. Thus, it may be that the conventional modern strategy of closed fishing seasons can be implemented during periods when most fishers have turned to nonfishing activities anyway.

Finally, it may be worthwhile for fishery officials to underscore the importance of any localized customs or beliefs that have the effect of limiting fishing effort or the demand for seafood. Public information campaigns to stir interest and sympathy for fishing peoples among members of the non-fishing public would be one way to accomplish this.

Changes in a fisheries management regime should always be implemented slowly. Every attempt should be made to discover, retain, or re-institute cultural practices and traditions that passively limit fishing effort and fish mortality. There should be no haste to impose gear restrictions or bring in new technologies. In particular managers should consider seriously the negative and often irreversible consequences that may ensue from development programs aimed at transforming fisheries from subsistence production to market production.

The log catamarans called *jangadas* of the northeast Brazilian coast are among the most primitive watercraft used in offshore fishing anywhere in the world today. Photo courtesy of Shepard Forman.

This fisherman's helper, who has just unloaded the fish, will carry them, hanging from an oar, to market, Paratinga, Brazil. Photo by F. Mattioli, courtesy of the FAO.

Fisherman throwing a cast net near Paratinga. Photo by F. Mattioli, courtesy of the FAO.

Shrimp trawlers under construction at a shipyard in Mazatlán, Mexico. The challenge of managing the world's fisheries is greatly compounded as ever more fishing vessels are added to fishing fleets around the world. Photo by J. R. McGoodwin.

Aboriginal oyster shell mound constructed ca. 1200 in south Sinaloa, Mexico—one of hundreds of such mounds that dot the region's coastal plain. The mound contains pre-Columbian pottery and other artifacts, suggesting that it may have served as a house platform. Harvesting marine resources on this scale may have required the organized efforts of a stratified society and prompted the need for rudimentary forms of fisheries management. Photo courtesy of Stuart D. Scott.

While industrialized fishing most often implies large-size fishing vessels, it also includes the fleets of smaller vessels, such as these shrimp trawlers at Guaymas, Mexico, that are organized and operated by a single corporate entity. Photo by J. R. McGoodwin.

A shrimp trap or weir near Escuinapa, Mexico. Although not technologically sophisticated, weirs such as these erected across strategic narrowings of an estuarine channel can be very productive. Similar weirs were used here in pre-Columbian times. Photo by J. R. McGoodwin.

Rural fisherman working just offshore, Pacific Mexico. This is the late Victor Osuna Tisnado (the Victor to whom I have dedicated the book), a poor rural fisherman who befriended me during my first fieldwork in Pacific Mexico, and who was lost at sea in a hurricane after he had found "better" work aboard a shrimp trawler out of Mazatlán. Photo courtesy of David B. Powell.

Small-scale fishermen hauling in a shark caught on a longline off the coast of Pacific Mexico. Although their boat is equipped with neither communications nor safety gear, these shark fishermen sometimes venture as far as 80 miles offshore. Photo by J. R. McGoodwin.

Fishing canoe, St. Lucia, West Indies. Except for the addition of an outboard motor, these canoes differ little from those used by the Carib Indians throughout the Caribbean Sea in pre-Columbian times. Photo by J. R. McGoodwin.

Small shrimp trawler in Galveston Bay, Texas. Individualistic small-scale fishers such as these collectively have a great impact on marine resources, and because of their dispersion and great numbers are often difficult to manage and control. In highly industrialized regions like this, they may also be plagued by problems of marine pollution and come into conflict with other competitors for coastal resources. Photo by J. R. McGoodwin.

Salishan and Sahaptian compound fish weir used by Native Americans to harvest salmon along the Pacific Northwest coast. Although technologically simple, large weirs such as these made possible such large catches that tribes zealously controlled the sites where they were located. Nowadays modern salmon-harvesting operations are conducted at many of the same sites. Drawn by and courtesy of Deward E. Walker, Jr.

Each year in Oregon, more than 25,000 people line the banks of the Sandy River east of Portland to dip for smelt, a small, migratory pan fish. The smelt run has become erratic over the past few years and occasionally does not occur in quantities large enough for successful dipping. Photo courtesy of the Oregon Department of Transportation.

Commercial fishing by Indian dip netters at Celilo Falls, Oregon. Overhead wires are for trams to take the dip netters from harvest sites to shore. Celilo Falls was an important fish-harvesting site long before European colonization. This picture was taken ca. 1950, before construction of the Dalles Dam inundated the falls. Photo courtesy of the Oregon Department of Transportation.

Facing page, above: Deckhands separate a purse seine into lead, web, and cork piles as a power block pulls the net from the water, Alaska. This is the most physically demanding task in salmon seining and crews take great pride in how fast they can retrieve their gear. Among these fishers strong aesthetic principles underlie the laying of each kind of gear pile. Photo courtesy of John B. Gatewood.

Facing page, below: Small-scale fleet of salmon purse seiners in Ketchikan, Alaska. Photo courtesy of John B. Gatewood.

Small-scale fishing boats and gear at Menemsha, Martha's Vineyard, Massachusetts. New England fishers are among the most staunchly independent and individualistic small-scale fishers anywhere in the world. Photo by J. R. McGoodwin.

A legendary skipper nicknamed "Binni" from the Vestman Islands, Iceland. For years one of the most successful skippers in the whole Icelandic fleet, he was once reported to have said to a less successful skipper, "You know why you don't catch fish? It's because you don't think like cod!" Photo courtesy of Sigurgeir Jónasson.

North Atlantic Icelandic fisherman bringing aboard a good catch of herring. A single haul of fish of this size is more characteristic of large-scale, industrialized modes of fishing, and would seldom be seen in any form of small-scale commercial fishing. Photo courtesy of the Icelandic Fishing Society.

Icelandic fishermen unloading their catch. Photo courtesy of Sigurgeir Jónasson.

Industrial-scale stern trawler at Skagen, a small port at Denmark's northernmost tip. Unlike its predecessors of only a decade or so ago, some of which exceeded 300 feet in length, this 144-foot ship is typical of the moderate-sized, multipurpose ships now plying the North Atlantic and North Sea fisheries. Photo by J. R. McGoodwin.

Fishing vessels in drydock for repairs at Skagen. In the literature dealing with fishers and the fishing industry, the important and integral role that chandlers, riggers, repair yards, fish processors, and marketers play in facilitating the overall fishing effort is often overlooked. In Skagen's local fishing industry, nearly seven times as many people are employed in support jobs ashore as work at sea. Photo by J. R. McGoodwin.

Fishing boats on the beach at Tossa de Mar, Costa Brava, Spain. This part of Spain's Mediterranean coast experienced such a rapid development of the tourist industry in recent years that the government was forced to set aside special areas on the beach where local fishers could continue to pull up their boats and land their catches. Photo by J. R. McGoodwin.

Fishermen rowing their catch ashore at Cap Lloc, Costa Brava. The lamps are used in night fishing to attract squid to the surface. Whereas in times past their catches supplied regional seafood markets, practically their entire catch is now sold to the resort hotels that abound in this popular tourist area. Photo courtesy of Oriol Pi-Sunyer.

EIGHT

Active Means of Indigenous Regulation

As we have seen, most purposeful indigenous management strategies attempt to regulate or limit access to fishing space, rather than levels of fishing *effort* (McCay 1978). Or put another way, "The object is not to protect or conserve the fish as much as to reserve the fish that are there for one's self" (Acheson 1981b: 281). In fact, asserting proprietary rights over fishing spaces and sanctioning the entry of only certain groups are probably the most commonly employed means of active and purposeful fisheries management used by local groups of fishers.

But other means are used as well. Indeed, some are analogous to those employed by modern fisheries managers and are congruent with modern conservation methods based on sound biological and economic principles. The question for this chapter, as for the last, is, are any of these indigenous management strategies worth incorporating into modern policies?

The Control of Fishing Rights and Space

The most popular technique for limiting fishing effort employed by modern fisheries managers is also the most common active strategy used by indigenous groups—the control of fishing space. This is not particularly surprising, since space is fixed and easier to identify than are marine resources, which even when sessile usually remain at least partially hidden.

That many fishing communities exhibit "clannishness" and a tendency to exclude outsiders has been so often noted that it is practically cliché in studies of fishing communities.[1] Indeed, many landlubbers practically ex-

---

[1] I place "clannishness" in quotation marks because I use it here in its colloquial sense, not in the anthropological sense of true kinship, implying lineally related relatives who claim descent from a common ancestor.

pect such behavior from most local, small-scale fishers. To name only a few of the better-documented examples, the phenomenon has been noted among small-scale oyster fishers in a rural community on Apalachicola Bay, Florida, where kinship ties define possible participants in the fishery (Rockwood 1973); in Newfoundland (Andersen 1979, 1982; K. O. Martin 1979); in Maine (Acheson 1979, 1988a); in Sweden (Löfgren 1982); in northwestern Washington State (Suttles 1974); and in Oceania (Johannes 1978).

It would be erroneous, however, to assume that such assertions of exclusive property rights are a relatively recent innovation in the fisheries, arising only in societies with well-defined institutions prescribing property ownership or in modern situations where the fisheries are characterized by depletion and chaotic competition. Though the earliest prehistoric, nomadic, hunting-and-gathering bands may not have asserted ownership rights over the vast hunting territories available to them, they probably asserted rights of privileged access to certain fishing sites that were relatively limited. And certainly with plant domestication and the shift from nomadic to more settled ways of life, highly productive fishing territories began to be regulated under increasingly explicit regimes prescribing proprietary rights—in much the same manner as garden plots, house sites, personal effects, and so forth (Andersen 1974).

In Japan, for example, coastal villages have for centuries asserted common property ownership claims over their traditional fishing grounds and striven to prevent entry by outsiders. Often the recognition and enforcement of these community ownership rights was administered through provincial, prefectural governments (see Befu 1981; Chang 1971; Matsuda 1972; Nishimura 1975; Norbeck 1954; and Ruddle and Akimichi, eds. 1984).

Indeed, Japan's contemporary local sea tenure systems have deep historical roots; many of them antedate even the feudal era, or Edo Period (1603–1867). Nishimura (1975), for instance, explains how primitive stone tidal weirs, the earliest and most primitive type of fishing technology used in Japan, "a man-made replica of the natural hollow found in a rock or coral reef which fills with water during the ebb-tide" (pp. 77–78), were owned in common until feudal lords took control of them and began not only regulating production but also requiring users to pay a tax for their right to fish.

These and other early Japanese fisheries management systems were codified by the Edo governments and still constitute important guidelines on which policies are formulated today (see Ruddle and Akimichi 1984: 6; Kalland 1984). Quite similar historical developments took place in the coastal fisheries of Okinawa, with legacies of these long-standing traditions still evident in intracommunity decision making as well as in fishers' assertions of territorial claims (Akimichi 1984; Akimichi and Ruddle 1984).

Similarly, in the Philippines and throughout most of Oceania prior to European colonization, chiefs, clans, communities, or families asserted ownership rights and exclusive prerogatives of administrative control over their important fisheries (see Blair and Robertson 1905 on the Philippines; and on Oceania, Johannes 1975, 1977, 1978; Klee, ed. 1980; Knudson 1970; Lessa 1966; Sahlins 1958; Sudo 1984).

The local assertion of exclusive property rights in fisheries has also been described among the highly advanced chiefdoms inhabiting the Pacific coast of northwest North America at the time of first contact by European explorers (see Drucker 1965; Hewes 1973; Suttles 1968). In what is now Washington State, individual Salish Indians enjoyed formal, inherited property rights over prime oyster beds and some fishing grounds (Suttles 1974). Similarly, in what is now northwestern California, the Yurok Indians asserted communal community rights over certain beaches and coastal fishing grounds (Beals and Hester 1974).

Other early examples of the active assertion of property rights have been surmised among various aboriginal chiefdoms living far to the south. In Pacific-coastal Mexico, prime sites for weirs, as well as particularly rich oyster beds, were apparently under the administrative control of chiefs as early as 1200 (S. D. Scott, ed. 1967–73).[2]

These examples may suggest that the practice was confined to simple societies, but of course it was not. As early as the thirteenth century in Britain various regional kingdoms are known to have asserted both rights of ownership and rights of administration over prime fisheries (Moore and Moore 1903). These kingdoms cannot be characterized as true monarchies

[2] Among many of these aboriginal groups, community rather than individual ownership of food-producing resources was the rule, and even when individuals did lay claim to such resources, their ownership was not absolute—certainly nothing like ownership as we conceive it. Instead, it was usually a sort of trusteeship granted by the community with the understanding that the individual had no right to alienate resources that were essential to the community's well-being. Because most of these societies had well-integrated means of social and economic control, only rarely did they allow their commonly owned property to suffer "the tragedy of the commons" (McEvoy 1986: 30). By asserting their common ownership against the claims of other competitors, they indicated that they had already learned what could happen if production was otherwise uncontrolled.

but were only complex societies that had achieved the level of political organization that cultural anthropologists associate with chiefdoms or nascent states. Similarly, in seventeenth-century New England throughout the Cape Cod region, the alewife catch was regulated by a town warden who specified certain days for fishing and permitted only town residents to participate in the fishery (Hay 1959).

Among today's new generation of professional fisheries managers, turning away from common property, open-access regimes to regimes that rely on property rights is often regarded as a novel and experimental approach to fisheries management. But, as we can see, this was an important means of active fisheries management practically from earliest recorded history. Moreover, in some parts of the world it managed to survive the spread of the doctrine of the freedom of the seas and remains in force today. In the Philippines, for example, some of the prime sites for fish corrals that were owned and administered by the Tagalog chiefs in precolonial times, then taken over by the Spanish authorities on behalf of the local communities, are now owned and administered by the adjacent municipal governments— a surprising persistence of traditional property rights in a country where most other coastal and offshore fishing grounds have long since become national common property resources (Spoehr 1980: 26). Similarly, many south coastal villages in India still assert property rights over particular fishing grounds, even though legally these fisheries are now deemed to be common property resources (K. Norr 1972). In various parts of Oceania, traditional community-based precolonial systems of marine resource management that confer property rights or prescribe limited entry are now being reinstituted in certain fisheries (Johannes 1978). Some of the peoples living around the North Atlantic rim have also managed to assert their control of their traditional fisheries. In parts of northern Norway, the assumption of rights to marine territories and resources through inheritance and marriage are still in effect (Brox 1964). And the artisanal hand-line fishers of Bermuda have developed a system of "positional tenure" over shallow-water fishing areas, particularly in areas that are customarily baited (Andersen 1976).

From this and countless other examples, past and present, there can be little doubt of the fundamental importance of proprietary rights as a means of management among local fishing peoples. It is a management tool that merits prime consideration in the formulation of new fisheries management policies, not only because the practice is so widespread, but also because it is congruent with many strategies currently being employed in fisheries management.

Political Activism

Just how fishers assert their rights over fisheries presents an interesting array of behaviors, ranging from concerted political action through the resort to extralegal means. Undoubtedly, political activism is one of the most potentially important legal methods for securing preferential rights to fishing spaces and limiting access to them. Unlike their mostly individualistic and apolitical predecessors, small-scale fishers are increasingly becoming politically active.

This statement, I am aware, flies in the face of the common assumption that because of their rugged individualism, small-scale fishers are not particularly inclined toward cooperative action. Moreover, at least two maritime anthropologists have argued that the linearity of coastlines geographically isolates such fishers from one another, making concerted action among them relatively difficult to achieve even when they are so inclined (Hewes 1948; Nishimura 1973). Even conceding that it is more difficult for small-scale fishers to come together than it is for industrial fishers, and further, that it is generally more difficult for all types of fishers to organize politically than it is for other occupational groups, the assumption that small-scale fishers are inherently apolitical still seems to me overdrawn. There is too much evidence to the contrary.

One prime example, the formation of fishing cooperatives for marketing purposes, is a familiar enough phenomenon around the world to need no embellishment here. But let us consider some of the numerous other noteworthy cases where small-scale fishers have shown the will to join forces in defense of their fisheries. Pi-Sunyer (1976) describes how a group of small-scale fishers in Spain managed to bar a yacht club from building on what was traditionally "their" beach. The importance of the beach to the local fishers was more symbolic, a matter of history and tradition, than it was economic at the time the controversy arose. Nevertheless, they took their case to the central government and won a judgment that reserved for them this important space.

In another example of considerably more strident political action, gill-net and drift-net fishers in North Yemen's Red Sea fisheries, after suffering gear destruction and resource depredations when trawlers entered their traditional fishing grounds, organized politically and got their central government to withdraw permission for the trawlers to operate there (Thomson 1980: 3).

Similarly, small-scale fishers in Indonesia were moved to political action after suffering severe conflicts with the entry and growth of fleets of trawlers and purse seiners in their traditional fishing grounds. Their strident protests, which involved not only legal political protests, but also rioting and violent acts against trawler fishers, were eventually influential in securing a government decree banning coastal trawling and proclaiming that small-scale fishers would be the priority group in the nation's fisheries development programs (Bailey 1986; Thomson 1980: 4).

These are not mere anecdotes or isolated examples. Throughout the world, as small-scale fishers are becoming less isolated with advances in communications and transportation, they are also becoming more politically active and aware. As competition for fisheries resources becomes more intense and the process of fisheries policy formulation more politicized, many of them have become sophisticated enough to participate more fully in the establishment of fisheries policies. Furthermore, in many developing nations the small-scale fishers' contribution to national food supplies is being more widely publicized, something that can only help them in their struggle for the enacting of laws that will limit access to their fishing grounds (see Bailey 1986; Charest 1979; Durrenberger and Pálsson 1987a,b; Johannes 1978; Osamu et al. 1979; Pinkerton 1988; Thomson 1980).

Extralegal Means

One of the most problematic ways local fishers assert proprietary rights and exclude "outsiders" is through resort to extralegal or illegal means. This occurs most often in common property, open-access fisheries and is particularly commonplace among small-scale fishers.

Methods for excluding outsiders range from the mere insistence on the rights of a preferred group to the exertion of pressure and even intimidation. Some extralegal methods may be comparatively benign: social ostracism, verbal abuse, misinformation, failure to render aid to boats that have broken down and are not in immediate danger. But there may also be attacks on people, the destruction or theft of gear, the burning of competitors' boats, and other forms of sabotage.

Local fishers' assertions of their right of exclusive access are illegal when they directly contradict a formally instituted legal regime in which a state or federal agency has been granted the prerogative of prescribing eligibility for entry to a fishery. Thus as Andersen (1982: 24) observes, the "development of public, legally enacted management schemes often does not stop the exercise of indigenous strategies for self-regulation, not even when the new legal managerial regimes render the indigenous strategies illegal."

At best, the extralegal methods employed by local fishers may duplicate certain management stipulations, so that their claims have little impact on the smooth execution of the legally instituted management regime. At worst, however, such methods may be in direct opposition to the legally instituted regime and may greatly confound its smooth operation, becoming a source of bitter and ongoing conflict. It is therefore of paramount importance that modern fisheries managers realistically anticipate the possibility that local fishers will develop extralegal means for controlling entry into their fisheries. Merely dismissing such behavior as illegal, and perhaps regarding it with disdain or contempt, will not make the behavior go away, nor will it help to establish a more appropriate management policy.

One of the best-documented examples of local fishers asserting extralegal fishing rights is provided by Acheson in his descriptions of the Maine lobster fishery in the United States (1972, 1975, 1979, 1982, 1988a). Maine lobstermen, he points out, have long resisted the state and federal governments' attempts to manage their fishery by trying to restrict access on their own. Essentially, this fishery is a common property, open-access resource. Various state laws regulate the inshore regions, and federal laws mainly the offshore regions. Theoretically, anybody wishing to fish there may do so, provided that the person obtains the necessary licenses and abides by the various state and federal laws. In practice, however, entry into the fishery by "outsiders" is actively discouraged by the locals.

Each fishing community has its own "harbor gang," community members who recognize each other's right to work in the fishery.[3] Such recognition is extended only to those who are born in the community and can claim long-term residence there, who have many relatives in the harbor gangs, and who begin fishing while young and do not start out too ambitiously. Each of these harbor gangs also asserts rights of exclusive access to particular fishing territories associated with its harbor. These territories are explicitly defined close to the home harbor but more ambiguously so the farther removed they are. At great distances from the home harbor, these claims may overlap the claims of harbor gangs from other coastal

[3] The word "gang" here does not carry the negative connotation we think of in connection with, say, an urban street gang, but is used in the sense found in the song, "Hail! Hail! the gang's all here," that is, to imply camaraderie (Acheson, pers. com. 1989).

communities. Interlopers who intrude into a harbor gang's territory risk reprisals of various types, including verbal or physical abuse, the release of their catches, and the destruction of their gear.

Interestingly, a study by J. A. Wilson (1980) indicates that when the price of lobster rises, these territorial enforcement systems break down as outsiders boldly invade the territories of the harbor gangs. The increasing incomes and rising expectations among all participants in the fishery result in a less-strident and more limited assertion of territorial claims. Just the opposite situation undoubtedly obtains when the lobster market falls. Thus the harbor gangs' territorial claims are not fixed by longstanding tradition but are instead responsive to changing economic conditions.

Similar extralegal assertions of exclusive fishing rights have been reported along Canada's Newfoundland coast (Andersen 1982). As in Maine, the sanctions levied against outsiders range from verbal abuse and "sheer aggressive dominance" to more extreme measures, such as cutting away the trawl lines of the interlopers.

The extralegal assertion of property rights may extend beyond just fishing territories. Other sorts of spaces important to fishers may be claimed as well. Among these, traditional beach landing sites and harbor spaces are probably the most important. On one occasion, small-scale fishers in Spain physically barred large vessels from landing their catches in an area of the town's beach that they had traditionally used (Pi-Sunyer 1977). Similarly, small-scale fishers from Goa, India, after suffering years of declining catches in their traditional coastal fishing grounds because of the incursion of industrial trawlers, eventually took matters into their own hands. Objecting to the presence of the large trawlers in the community harbor, they set two of them on fire (Osamu et al. 1979).

I have personally observed many instances of the extralegal assertion of both rights of exclusive access and trade and commerce rights among small-scale fishers in rural Pacific Mexico. Local fishers often sabotage or steal gear set by fishers from neighboring villages in territories that they illegally assert to be their own preserve. Moreover, while federal laws have set aside many traditional fishing grounds for the exclusive use of local fishing cooperatives, impoverished nonmembers openly fish in those territories and more than once have committed violent acts against local cooperative officials, as well as government agents dispatched to the area to enforce the law (McGoodwin 1980, 1987).

Furthermore, prior to Mexico's assertion of its 200-mile EEZ, foreign longliners, mainly Japanese, legally set their lines in international waters close to Mexico's Pacific coast. These waters were also worked by Mexican shark fishers, who often showed me Japanese fishing gear they had taken for themselves. It was appropriate to steal this gear, they insisted, because they had found it in "their" waters. They also insisted that taking the gear denied catches to the Japanese, which helped to conserve shark stocks for themselves. Of course, by stealing the gear the Mexican fishers also avoided considerable expenditures on gear replacement.

In the fishing villages along this part of the Mexican Pacific coast, the entry of new middlemen and other marketers is also vigorously discouraged. Two main means, one illegal, the other not, are employed. The illegal strategy involves local middlemen encouraging their dependent clients to subject newcomers who try to set themselves up as competitors to verbal abuse, sabotage, and sometimes violence. The legal strategy is related. Local middlemen try to tie up the catches of local fishers through the age-old system of debt peonage, shutting out other, more recently arrived middlemen.

Biological Control

Many of the control practices indigenous peoples employ have analogs in modern fisheries management strategies that are rooted in biological and conservationist concerns. Though these practices appear to be motivated by the same concerns, closer examination reveals that some may not have been instituted for such reasons at all. Still, the question of motivation is not so important for modern fisheries management as the question of a practice's potential for controlling or limiting fishing effort.

Limiting access, which is an indirect way of reducing fish mortality, is still the most common strategy employed in indigenous regulatory schemes. However, even where this is the cornerstone of an indigenous management regime, it is usually augmented by other means, many of which are also analogous to conservationist measures employed in modern fisheries management. These other means include curtailing effort once stocks are seen as being unduly pressured, imposing catch limits, returning fry to the water, discouraging overzealousness by fellow fishers, observing closed seasons to ensure the survival of spawning stocks, and improving the marine ecosystem.

Indigenous means of fisheries management employed by preindustrial and premodern peoples that seem to have close parallels with modern biologically based strategies have been widely reported. Although the record is somewhat anecdotal and does not contain an abundance of examples, there is no doubt that many indigenous peoples understood the rudimentary biological requirements of their fisheries. Indeed, Johannes (1978: 352) insists that "almost every basic fisheries conservation measure developed in the West was in use in the tropical Pacific centuries ago." The list includes banning fishing during spawning periods, allowing a portion of the catch to escape, deliberately not taking all the available stock of certain species, holding excess catches in enclosures until needed, prohibiting the taking of small individuals, and restricting the number of traps in an area. In parts of Oceania, master fishers were even assigned to act as "conservation officers" and "fishery ecologists," with the power to authorize and organize fishing parties, lead communal fishing efforts, and restrain fishing effort in order to conserve resources (Klee, ed. 1980).

Though the active imposition of gear restrictions for the express purpose of limiting catches—probably the oldest form of fisheries regulation in the West (Johannes 1978: 355)—seems to have been rarely employed, a few instances have been reported. The Cree Indians of northern Canada, for example, intentionally use gill nets that reach only to limited depths so as to permit some adult fish to escape. They also use certain minimum net-mesh sizes to permit immature fish to escape (Berkes 1977). They actively regulate their fisheries in other ways, too, as we shall shortly see. For the most part, however, fishers use the most effective fishing gear they can get their hands on.

The conservationist picture is mixed at best. The principle was clearly understood, if not practiced in an aquatic ecosystem, by the peasant farmers of ancient Asia and medieval Europe, who carefully added plant trimmings, animal manure, and other biodegradable materials to their aquaculture ponds in order to supply them with vital nutrients (W. A. Johnson 1980: 179). Many scholars have pointed to the multiple benefits of such practices, which not only helped solve the problem of disposing of noxious garbage and other refuse, but also significantly increased food production. Instances of ecosystem enhancement are harder to document among preindustrial and premodern maritime people. But at the very least there must have been cases where people recognized that they could increase the productivity of their marine ecosystems by physically altering them through canalization, dredging, and the like.

What of efforts by preindustrial and premodern peoples to curtail fishing out of specifically biological concerns? Such efforts have been widely reported from many parts of the world. "The King's Gap"—a requirement in thirteenth-century Britain that salmon fishers in the county of Cumberland space their riverine nets far enough apart to allow some of the spawning fish to escape as they headed upstream falls into this category (Moore and Moore 1903). So may the regulations imposed by various chiefdoms that depended on the salmon fisheries of Pacific North America. Limits were placed on the times of day and the seasons in which the fish could be harvested; catch limits or quotas based on need were observed; and traps had to be removed at certain times so the salmon could proceed upstream. These measures were regularly used at Celilo, near the boundaries of the present-day states of Washington and Oregon, and at Kettle Falls, near Spokane (Deward E. Walker, Jr., pers. com. 1983). *salmon*

But whether these strategies were truly designed to sustain the resource is problematical. Gordon W. Hewes suggests that the main motivation for allowing salmon to escape was to maintain alliances and forestall the aggression of people living upstream (pers. com. 1983). And many of these chiefdoms utterly wasted huge quantities of their marine resources in their renowned potlatch ceremonies. In these elaborate rites, chiefs displayed and destroyed large quantities of surplus wealth, including huge amounts of seafood, in order to attract and hold allies, thereby solidifying their political power and prestige.

So the evidence is mixed. It may be that even though many of these peoples sought to reduce fishing effort, they were in fact anti-conservationist and wasteful in the extreme. The intent may have been to conserve marine foods only so that they could be wasted later on. A case like this suggests that we should be careful not to impute a conservationist purpose to indigenous practices just because they appear to be analogous to modern strategies. It would be better to withhold that judgment until a society's overall pattern of resource exploitation is examined.

We are on somewhat safer ground when it comes to modern fishers. Groups in many areas have been seen to employ conservationist measures in order to sustain stocks, at least so long as access to the fishery is limited, and they are not experiencing dwindling yields or chaotic competition. Artisanal fishers in Honduras, for example, may return small fish to the water so they can grow to larger size (Jan Peter Johnson, pers. com. 1982). And small-scale lobster fishers in Maine may verbally abuse fellow harbor gang members for their overzealous attempts to increase production by fishing in bad weather or setting out too many traps. Many of the Maine lobstermen have also avoided adopting more effective gear such as metal lobster pots, for fear that these might lead to overharvesting (Acheson 1982).

Similarly, Berleant-Schiller (1982) notes that the lobster divers of Barbuda, in the Caribbean Sea, spare gravid females and also turn away from fishing when declining yields indicate their prey is being overharvested. She points out that the local fishers are able to meet subsistence needs and redistribute their incomes among the island's populace, but do not earn enough money to finance higher-yield modes of fishing. However, she warns, "this working balance could easily be upset by mechanization aimed at expanding production" (p. 134).

McCay (1981a: 4) reports instances of voluntary restraint among the clam fishers she studied in New Jersey, lest they (in their own words) "send the clams to the same fate as the American buffalo." They also selectively thin the stocks for a better harvest. But then, as she notes, these are commercial fishers, and their aim is not only to sustain stocks, but also to support or drive up the market price of their catches.

Berkes (1977) provides a particularly interesting case study of the Cree Indians living around James Bay in northern Canada. Though this group is sufficiently large and well equipped to overharvest its fishery, the fishers exercise self-restraint, and as a result enjoy sustained high yields year after year. The Cree fishery, Berkes notes, is "characterized by a high degree of order," thanks in part to limits being placed on the intensity of fishing effort and even the types of gear employed. These restraints arise from a social ethic that calls for catching only the amount of fish required for immediate needs and discourages any overzealous effort. But the Cree do not stop there. They also exploit only some of the good fishing spots available to them in order to leave several unfished sanctuaries for their mainly targeted stocks. Moreover, as mentioned earlier, they intentionally use gill nets of a size and mesh that permit the escape of immature fish and some adults as well.

Though the Cree fishers have thus far maintained a working balance between themselves and the marine ecosystem they exploit—even as their population has grown and they have adopted more modern fishing gear—Berkes (1977: 306) warns that, "with population growth and/or the provision of an incentive to accumulate a surplus catch (such as external market demand), the situation could be different in the future." Thus it is worth emphasizing that this smooth-running local management regime appears to be mainly effective because outside market forces have not yet impinged on it.

Information Management

As we saw in the previous chapter, information management can be both a passive and an active means of reducing fishing effort—passive for those fishers shut off from advantageous information, active when used by those who possess it. A number of researchers have observed that information management may serve to separate fishers spatially, exclude or discourage the entry of new fishers in a fishery, or otherwise serve to reduce overall fishing effort (Acheson 1981b; Forman 1967, 1970; McCay 1981a). For instance, fishers within a given fishery often form alliances such as "code groups" whose members share information through radio communications and seek to deny it to fishers outside the group (Orbach 1977; Stuster 1976, 1978).

Various writers have suggested that information management, particularly when it involves keeping good fishing spots secret, confers a temporary, de facto property right on those in the know (Andersen 1972; Forman 1967, 1970; Löfgren 1972; Stuster 1978). Andersen (1972) in fact calls this "informational capital." The various means of information management he found employed by the patrilineally related groups of small-scale fishers in Newfoundland included not just hoarding information about especially good fishing locations, but also giving misleading information to other fishers—mainly through understatement—in order to lead them away from the desirable spots or cause them to waste time and effort. Another tactic was to avoid landing the full catch at any one time or place in order to disguise its actual extent.

This sort of information management is comprehensively described by Shepard Forman in two well-known studies of the raft fishers along the Brazilian coast (1967, 1970). These fishers all view the fishery as an open-access common property resource, and information about good fishing grounds is more or less common knowledge. But information about particularly good spots within these grounds is a highly guarded secret. The common knowledge of good grounds, Forman argues, guarantees at least moderate success for everybody, while the secret knowledge about particularly good spots prevents excessive competition there, reducing the possibility of their becoming overfished. Something of the same practice is reported among the small-scale fishers of Arembepe, Brazil (Kottak 1966).

Discussing the value of indigenous information management strategies of still another Brazilian group, Cordell (1974) describes an esoteric sys-

tem of knowledge employed by impoverished inshore fishers in an estuarine region. He sees their highly specialized knowledge about how to
exploit various microenvironments during different seasons as tending to
space out their fishing effort, but is not certain that this specialized knowledge played a decisive role in preventing overharvesting in the past, since
most local fishers eventually learn the system.

In my research in Pacific Mexico, I found a marked contextual difference in information management among the region's small-scale fishers.
Basically, I observed that fishers tended to be secretive and misleading
when they worked inshore waters, and that there was an almost opposite
sort of behavior when they fished farther out (McGoodwin 1979). The
disparity undoubtedly results from different perceptions of the overall
availability of marine resources in the two environments.

My observations in Mexico parallel those of Cordell (1978) for a very
similar group of fishers in northeastern Brazil. Because traditional inshore
fisheries are "fixed-territorial" in nature, Cordell observes, specific spaces
are more likely to be regulated by the local people there than in the offshore waters.

Orbach (1977: 104) provides some particularly interesting observations about the formation of code groups and the use of information management in various types of fisheries. "There are certain fisheries," he
states, "where attempting to use information strategies to find the 'best'
areas does not seem to yield more than a random fishing pattern does
(Dickie 1969), and others where a reduction in the information potential
among the fleet seems to affect the catch dramatically (Wadel 1972)." The
skippers of the San Diego tuna seiners that he crewed on between 1973
and 1975 actively joined code groups, but in their radio communications
with other vessels, they deliberately limited the information they shared
even with their own group. Curiously, Orbach saw no evidence that their
attempt to mislead the other skippers ultimately made much difference in
their overall catches. So why did they do it?

Orbach's answer is that they did it for at least three reasons. First, because among these fishers participation in the information management
game *seemed* to be efficacious, even though this was only an assumption.
Second, this was a conspicuous means for the skippers—who were the
sole controllers of the information that was exchanged—to underscore
their special status as skippers; for them, radio communications were a
means of social display. And third, the back-and-forth by radio helped to
relieve some of the boredom of the long periods at sea (pp. 104–5).

More generally, it was clear to Orbach that these communications were undertaken purely out of a wish to be sociable, a feeling that is heightened in the limited and risky context of ocean fishing. So these communications were made in recognition that despite their competition, all these fishers were engaged in a common activity and might have to call on one another for help in times of distress.

In the end, the skippers of fishing vessels at sea must all confront what I call "the skipper's paradox" in communicating with competing vessels. They must appear to be helpful if they expect to gain any information in return, but at the same time, they must mislead other skippers so they will not swarm to a hot spot and reduce their own catch. In essence, they are involved in a mini-max game in which the rule is to maximize one's gains of valuable information and minimize one's losses of it. This paradox, of course, ultimately derives from the requirement that these skippers be socially participant cooperators and competitors at the same time. Not participating, given the perils of ocean fishing, does not seem to be a strategy with much to offer.

In general, studies of fishers support two conclusions: that the more vulnerable fishers feel a given marine resource is, the more they will tend to hoard or control information about it; and that even when fishers are engaged in harvesting an abundant and underutilized stock, they will usually show marked proclivities for communicating with one another while at sea (see Gatewood 1987; Orth 1987).

Skill Differences

My inclusion of skill differences as an active or purposeful means of management is somewhat arbitrary. The behavior of a master fisher who is so skillful that he consistently outproduces his peers and who takes care not to reveal the secrets of his success can be construed as an active, purposeful management strategy. But the lack of know-how that prevents the fisher's peers from keeping up can obviously only be construed as a passive means of limiting productivity and fish mortality. The reluctance of fishers to share this kind of information has been widely reported (Acheson 1977; Andersen 1980; Drucker 1965; Faris 1966; Forman 1970; Nemec 1972; Paine 1957).

Fishing peoples themselves use what we may term various folk models for explaining why some members of their group are more successful than others. Most models emphasize differences in skill, fishing tactics, or initiative; luck; or magical or psychic attributes of one sort or another, such as foresight or intuition (Acheson 1981b; Barth 1966; Firth 1946: 99; Orbach 1977: 82).[4]

The existence of significant differences in skill among an otherwise homogeneous group of fishers has important implications for modern fisheries management. It is possible that in certain fisheries these may play far more of a role in limiting productivity than differences in effort as such. If so, the matter assumes particular importance in fisheries worked by people who are otherwise peers and who all employ essentially the same types of fishing vessels and gear (see Acheson 1977; Andersen 1972; Baks and Postel-Coster 1977; Cordell 1974; Forman 1967; Heath 1976; Jakobsson 1964; Norr and Norr 1978; Poggie 1979; Wadel 1972).

An astounding observation I made one summer during my early field studies in Pacific Mexico was the vast difference in catch sizes among several dozen handline fishers from the same village who all fished in one small spot just off the mouth of an estuary. Over the several days that I went out with them, they used identical bait and gear. A few caught twenty to thirty fish a day; others caught only two or three. As one of the master fishers explained, his success was due to his experience, which had taught him precisely how and when to set the hook as the finicky fish nibbled on the bait. But even after he had thoroughly explained his technique to me and I put it into practice, I was able to improve my overall catch only slightly.

In many fisheries the boats of certain captains are often claimed to yield exceptionally large catches season after season. In the English-speaking world, and especially in the North Atlantic fisheries, particularly successful boat captains are referred to as "highliners" or "fish killers," and their putatively greater success is often referred to as the "skipper effect." But some researchers seriously question whether there is such a phenomenon and argue that studies have not brought forward convincing longitudinal data to show that some captains consistently outperform others who bring similar levels of experience and fishing gear to their fishing efforts. Ascribing particular success to certain skippers, the critics say, arises mostly out of local myths and societal needs, not from any hard evidence of better-

---

[4] Luck is of course an important, deeply symbolic, and ubiquitous concept in the consciousness of fishers, used to explain in-group differences in catches by fishing peoples from such disparate places as Sri Lanka (Alexander 1977: 238), Newfoundland (Stiles 1972: 41), France (Jorion 1976), the Shetlands (Byron 1988), Sweden (Löfgren 1977), and the Trobriand Islands (Malinowski 1954), to name only a few.

than-average fishing skills. They especially fault most of the studies for not having adequately controlled for differences in boat size or fishing gear.[5]

For several scholars, the tendency to endow some skippers with phenomenal success serves social ends. In a study of fishers on Burra Isle in the Shetland Islands, a small community of only around 800 people, Reginald Byron (1988) proposes that by ascribing the success of certain skippers to "luck," the community legitimizes the leadership of those skippers, and, more important, has found "a diplomatic way of expressing distinctions between individuals and groups" (p. 3). He concludes that "explanations in terms of luck are a tactfully neutral way of speaking about social differences" (p. 14). Gísli Pálsson (1988) could find no statistical relationship between success rates and the fishing behavior of various skippers in southwest Iceland. Like Byron, he concludes that the popular accounts of outstanding success in Icelandic fishing villages merely serve social ends and do not reflect real differences among skippers. In his view, folk ideologies among Icelandic fishers do not have the long-term historical continuity that many have assumed, but are invented to meet certain social and economic needs. Before the modern era, he notes, when Icelandic fishing was still mostly a part-time occupation for subsistence, not the highly competitive commercial endeavor that it is today, ideologies accounting for the skipper effect and other aspects of differential success rates were nonexistent. "The rationality of the Icelandic notion of skipperhood," he states, "is largely located within the realm of social relations," that is, as modern responses to the social and economic exigencies Icelandic fishers now face (p. 25).[6] This is consistent with C. Alexander Goodlad's (1972) finding that when modernization brought changes in fishing tactics and fishing gear to the Shetland herring fisheries, folk ideologies about fishing changed too.

But Thorolfur Thorlindsson (1988) disagrees. His study of the skipper effect, controlled for all the above-mentioned variables, presents impressive statistical evidence for some boats captained by particular skippers outproducing others, frequently catching "four or five times more than an average boat during the summer herring season in Iceland" (p. 199). "The skipper effect," he concludes, "is not a 'folk myth' with no empirical

[5] The skipper effect is debated in Byron (1988); Durrenberger and Pálsson (1983, 1985, 1986); Gatewood (1984); McNabb (1985); Pálsson (1982, 1988); Pálsson and Durrenberger (1982, 1983, 1990); and Thorlindsson (1988).

[6] Gatewood (1983) notes, in this connection, how greatly skippers in certain fisheries are constrained by the social contexts in which they work, rather than by conventional notions of "rational" behavior, when making decisions on how and where to fish.

support in reality. It is supported by both statistical and qualitative evidence" (p. 210).

Though the evidence seems inconclusive, I am satisfied that there are cases where individuals in otherwise homogeneous groups of fishers employing similar or identical gear consistently bring in significantly higher catches than their peers. Since they presumably also bring about relatively greater fishing mortality, their particular attributes and behavior may have important implications for fisheries management policies. This phenomenon clearly merits further study.

Etiquette

Taking turns at fishing spots, permitting the first fishers to reach a spot to fish there first, and other forms of etiquette are another effective indigenous method for actively regulating access to fishing space (see Alexander 1977; Berkes 1987; Britan 1979; Catarinussi 1973; Cordell 1978; Orbach 1977). Etiquette systems are most effective among homogeneous groups of fishers from the same community; they tend to break down when dissimilar and unrelated groups of fishers enter a fishery. Yet in his study of small-scale beach-seine fishers in southern Sri Lanka, Paul Alexander (1977) describes a rather elaborate system in which competing groups take turns deploying their seines at a limited number of good spots. While these fishers regard their fishery as an open-access, common property resource, they are aware of the chaotic competition that would ensue if they all tried to work these spots simultaneously. However, though this system, described by Alexander as based on "the moral principle . . of equal opportunity" (p. 240), succeeds in preventing or at least minimizing cutthroat and chaotic competition, it has not succeeded in preventing overcapitalization.

Cordell's (1978) small-scale net fishers also observe a complex system of etiquette in their crowded inshore Brazilian fishery. Essentially, they assert temporary property rights over prime fishing spots by spacing themselves, if possible, or by honoring the principle of first come, first served. If a number of fishers arrive simultaneously at a good site, they draw lots to determine the order in which they will shoot their nets. Both Fikret Berkes and Raoul Andersen found similar systems operating in the groups they studied. "The Cree fishery can hardly be characterized as a free-for-all," Berkes (1987: 74) states. "Fishing operations are orderly [and] there is a code of ethics that essentially provides for respect for other fishermen and for the fish." And according to Andersen (1982), the small-scale traditional fishers of Newfoundland had "gentlemen's agreements" and also held drawings for named sites for placing cod traps.

Some Lessons for Modern Fisheries Management

In this chapter we saw that the most common way local fishers actively regulate fishing effort is by asserting some sort of proprietary rights to their fisheries, usually by limiting access to fishing spaces. It is an ancient practice and is still employed. There is a naturalness in this practice that argues favorably for its potential as a key component of future fisheries management policies. This seems all the more true now that the trend has turned toward privatization, unburdening many fisheries managers of the need to manage their domains as common property, open-access resources. Moreover, experience shows that when fishers enjoy what amounts to property rights, they usually voluntarily restrain their fishing effort. This in turn reduces the cost of monitoring and enforcement.

Instituting private property rights should also be considered a prime means for protecting fishers from outside competitors, as well as from other external threats that may erode the health of their fisheries, like the development of tourism and recreational fishing. Securing property rights for fishing peoples would put them in a position to participate in such developments instead of being helplessly victimized by them.

property rights

In most cases, rights of access or rights over marine resources and marine properties should be conferred on fishing communities, not on individual fishers, particularly when the communities are fairly homogeneous. The institution of territorial use rights in fisheries (TURFS), for example, a policy I will discuss in Part Three, has many parallels in both indigenous and modern management regimes and would be a particularly appropriate strategy in fisheries having sedentary and semisedentary resources. When such rights are instituted in a modern management regime, the fundamental principles in deciding who will qualify should be historical involvement in the fishery, the maximization of overall social benefits, and distributive equity.

distribution of rights

Local fishers are most likely to assert some sort of proprietary rights, as well as to engage in illegal activities, in highly pressured fisheries. In these situations policymakers should sympathetically explore the reasons that prompt this activism or illegal behavior and then attempt to revise the management policy in a way that will at least mitigate the problem, and better still, eliminate it.

As we have seen, it is a mistake to assume that traditional fishing societies rarely employ the kind of conservationist strategies now favored by modern fisheries managers. Several recent ethnographic studies suggest otherwise. Where it can be shown that local groups of small-scale fishers do employ such strategies, and that these are efficacious in limiting fish mortality, those strategies should be incorporated to the fullest extent possible into the fisheries' management policies.

As mentioned in the previous chapter, making the hoarding of information about fishing spots or techniques part of a management policy would seem ill-advised for humanitarian reasons. Permitting such differences to persist through official sanction would perpetuate socioeconomic inequality and inefficiency.

Local systems of etiquette should be considered for incorporation primarily or only where they exist among homogeneous groups that are not subject to intense competition from outsiders. They seem to work best in fisheries where the local norms prescribing appropriate behavior have had much time to simmer and steep. But some caution is in order here, because though such systems do sometimes limit fishing effort and fish mortality, most do not prevent overcapitalization. Instituting local customs of etiquette in modern fisheries management regimes seems appropriate mainly when fish are plentiful but fishing space is not, a rather uncommon situation. Moreover, once competitive pressures increase, and particularly when outsiders who do not honor the local group's long-standing traditions enter a fishery, a management regime relying on institutionalized etiquette will break down.

In sum, wherever any of the indigenous management strategies discussed in this and the previous chapter can be determined to be effective in reducing the mortality of marine resources, minimizing conflicts, effecting economies of managerial effort, and so forth, policymakers should capitalize on them in devising their fishery's management regime.

PART III

Fisheries Management Now and in the Future

We have here but five loaves, and two fishes.

Matthew 14:17

NINE

Needs and Problems

Few domains of natural resource management are more complicated than the fisheries. Diverse types of producers with diverse aims, as well as the mostly hidden character of the resources, make management particularly difficult. It is not unusual in one fishery to find tribal peoples fishing for subsistence, small commercial operators running family businesses, corporately owned trawlers from foreign ports, recreational fishers, diving enthusiasts, and so forth—all stridently asserting the primacy of their particular interests. Meanwhile, the fishery's marine resource populations may also be undergoing changes both as a result of all this activity and independent of it, some of which may not be detected until well into the future. Fisheries managers are therefore challenged with the complex task of maintaining resources at acceptable levels for all interested parties, balancing access and regulating use in a way that will achieve the most good for the most people.

Concerns over the depletion of fish stocks around the world, foreign incursions into domestic fisheries, and an appreciation of the inadequacy of the various management regimes in force have prompted nearly all the coastal nations of the world to reconsider their fisheries policies. With the institution of 200-mile EEZs, these nations are now faced not only with new opportunities, but also with new responsibilities. Unless these new management challenges are met, marine resources will continue to dwindle, competition and contention over them will become more heated and complicated, and the fisheries will fail to produce the great social and economic benefits they typically yield when maintained in a healthy state.

Ultimately, fisheries managers and policymakers must address the following general concerns if their regimes are to be effective: (1) human needs, values, and social equity; (2) biological conservation and resource productivity; (3) economic productivity and efficiency; (4) administrative feasibility; and (5) political acceptability (FAO 1983; Beddington and Rettig 1984). Account must also be taken, of course, of specific problems posed in individual fisheries around the world. Let us examine each of these items in more detail.

Human Needs, Values, and Social Equity

"Basic human needs" surely embraces a lot more than just food, clothing, and shelter, but when we ask what these needs are—what is essential to ensure "the good life"—we are likely to get a different opinion from nearly every quarter. The same goes for such abstractions as values and social equity. Safe to say, these matters have been the subject of a vast literature that far transcends mere concern for the fisheries. Yet the newly stated goal of optimizing social yields (OSY), while also maximizing sustainable yields (MSY) and economic yields (MEY), requires fisheries managers and policymakers to make decisions that must address these broad concerns, and usually on an ongoing basis.

Take social equity, for example. To some fisheries managers, this means making sure that fishers receive an equitable distribution of the catch or income from a fishery. To others, it means no more than ensuring equitable access. To still others, it means a whole host of other things. In fact, deciding which among many equity concerns should be paramount is probably the most difficult task fisheries managers and policymakers face.

How, for instance, can a policy balance the needs and desires of commercial and recreational fishers, or small-scale and large-scale industrialized fishers? Many have contended that comparing these groups is like comparing apples and oranges. Moreover, how can a fisheries management policy balance the often opposed interests of producers, marketers, and financial institutions? What are managers to do when they realize that the only way to prevent a fishery's total collapse is to curtail further fishing effort, while at the same time the fishers in their domain are stressing the importance of maintaining traditional fishing methods, their heritage and cultural identity, and freedom from restraint or other managerial encumbrances? What should be done when a policy satisfies most of the participants in a fishery but negatively affects the national economy and thus the general welfare of a nation's nonfishing populace? These questions only hint at the complexities with which fisheries managers and policy-makers must deal.

A people's heritage and preferred lifestyle, business profitability, recreation and leisure, overall national welfare—there is no manual fisheries managers can turn to that will tell them which of these should be primary, and which subsidiary. Moreover, none of these values can be considered apart from their consequences on others. Thus, whose profitability, say, should the manager and policymaker be most concerned about? What group or groups should be favored, and to what degree? Who should make the crucial decisions that determine who prospers and who goes broke?

Because so few management agencies rely on the expertise of fishers themselves, as well as that of cultural anthropologists, sociologists, and other social scientists and policy analysts, it is easy to understand why fisheries management problems have grown more complicated and contentious in recent years. Yes, overfishing is certainly the main problem, but it has just as certainly been exacerbated by inappropriate approaches to management and policies. Because most fisheries managers are still mainly specialists in fisheries biology, fisheries economics, or fisheries technology, few are equipped to evaluate human needs, values, and equity concerns in any but the most rudimentary way. Moreover, in the final phases of establishing fisheries policies, political pressure is usually as decisive as bio-economic concerns, a circumstance in which the advocates for one side or another more often confound the policy-making process than inform it. Balancing the interests of various competitors is also often unduly influenced by the prevailing ideology of whatever governmental regime is in power. Thus in many developing nations where preference was once shown for subsistence-oriented small-scale fishers, the policy was radically reversed with a change of political regime.

Practically speaking, one of the most difficult tasks facing fisheries policymakers and managers is to decide among alternative management strategies that will have the effect of improving the position of one group or class of fishers at the expense of another. Because it is not politic to admit to favoritism, preferential treatment is usually discovered only after its effects have become obvious or after it may be inferred through later analysis of fisheries regulations and policies. Few policymakers will publicly admit to showing favoritism, even though most will privately concede that favoritism is implicit in many of their decisions. The result is seen everywhere. Some nations may favor fishers who operate their own vessels over those working for companies or corporations. Others may favor near-shore and inshore fishers over offshore and distant-water fishers, drawing sometimes arbitrary distinctions between these groups. Still others may favor fishers from certain ports or regions or from certain ethnic groups. And, as previously mentioned, some favor full-time fishers, with severe re-percussions for part-timers. Some management regimes favor primary producers over processors and other middlemen; some take the opposite tack. Likewise, recreational fishers may be either favored or restrained, de-pending on the political and economic climate in the nation where they are found.

Unfortunately, the crucial policy decisions that result in favoritism are too often made on a "squeaking wheel gets the grease" basis. Moreover, as Beddington and Rettig (1984: 6–7) observe, "Even when one group is given a higher status . . . there is still the question of how much to favour that group." Thus, they wonder, even when official policy stresses that the interests of artisanal fishers should take precedence over the interests of the industrial fleet, "must you completely satisfy the artisanal group before providing any opportunity to the industrial sector?"

The more developed and pressured a fishery is, the more social equity concerns will occupy fisheries managers and policymakers, because it is in fisheries operating near bioeconomic equilibrium that we usually see the most extreme conflict, ill-will, and even violence.

Achieving full employment is also a knotty problem for policymakers who wish to promote human welfare and social equity. They may be under pressure, for example, from those who maintain that in the many small rural fishing communities where few alternative forms of employment are locally available, overfishing is preferable to unemployment. And yet, when carried to an extreme, a policy stressing the maximization of em-ployment might not only bring about the utter collapse of a fishery, but also have unfavorable consequences for the national economy and the na-tion's nonfishing populace. Somehow an appropriate balance must be struck.

Complex as the problems of human needs, values, and equity inevitably are, they are not insuperable. But solving them requires bringing all inter-ested parties more fully into the policy-making process. It also requires greater emphasis than we have seen till now on policy studies of the fish-eries. One important step in this direction is the founding of policy re-search programs like the Marine Policy and Ocean Management Program of the Woods Hole Oceanographic Institution. WHOI, which was founded in 1930 at Woods Hole, Massachusetts, supports the research of anthropologists, sociologists, political scientists, economists, lawyers, and others interested in analyzing fisheries management policies.

Biological Conservation and Resource Productivity

Maintaining the biological productivity of marine resources at high levels must remain of primary concern to fisheries managers, but this alone cannot address the economic and social problems that almost inevitably arise in modern fisheries, particularly those that are common property, open-access resources. So while a few policymakers still favor managing a fishery for MSY and otherwise letting market forces determine its economic yield, the unfavorable consequences that often follow in modern-world fisheries are usually sufficient to force even the most staunch free market advocates to curb their fervor. As Bell (1978: 339) puts it, "Even the most conservative economist, one who believes in minimum government participation in the free marketplace, will admit that Adam Smith's 'invisible hand' will result in market failure in the form of overcapitalization and eventually the demise of the fishery resource."

failure of MSY

Managing for MSY would avoid disastrous resource declines in most open-access fisheries, but any modern fishery managed solely for this objective would almost inevitably become overdeveloped, and then its fishers would suffer all the associated ills: excessive capital costs, intense competition, declining yields per unit of effort, and the utter dissipation of the economic rent in the fishery. This has already been the experience in many developed countries where fisheries policies unduly emphasized mere resource conservation, an experience that provides a clear warning to developing nations tempted to follow the same course.

Economic Productivity and Efficiency

Although most fisheries today are still managed as common property, open-access regimes, this is slowly changing. Moreover, as Bell (1978: 355–56) notes, many of these fisheries are now struggling to address four distinct types of market failures: (1) overcapitalization and overfishing; (2) environmental degradation by polluters resulting from the careless, uncontrolled, and indiscriminate use of common property water resources;

failure of common property, open access

(3) the misallocation of marine resources among commercial and recreational users; and (4) an inequitable subsidization of various producers designed to give them the competitive edge in the world market.

The first problem, overcapacity, is being addressed by the gradual erosion or dismantling of common property, open-access regimes. But it is still no small matter, as we will shortly see. Headway is being made against the second problem, water pollution, in some of the more developed countries. Pollution continues practically unabated in many developing nations, however, and overall it still poses a considerable long-term threat for the whole planet. So far it has been politically difficult, if not impossible, to make polluters pay for any reduction in economic rent they bring about in a fishery.

The third problem, the misallocation of resources among commercial and recreational users, is being addressed in various ways. In a few countries, a growing recognition of the impact on fish mortality caused by the recreational sector has prompted the collection of taxes and license fees, which are returned to the fisheries. But necessary change is often difficult to bring about. For example, even where it has been shown that recreational use has a lower impact on resource mortality, provides a wider array of human satisfactions, and ultimately makes a higher contribution to the economy than commercial fishing, politicians have been reluctant to shift fisheries policies against commercial fishing interests.

Bell's final problem, the subsidizing of fishers who compete in the world market, remains intractable for the moment at least. For some nations subsidization is as relevant in the domestic sphere as in the international sphere.

To return to the matter of overcapacity, since the turn of the century this has often been cited as the most dire problem in open-access regimes. As competitors in such fisheries acquire ever more powerful and productive vessels and gear, fisheries managers are increasingly forced to shorten fishing seasons, causing considerable disruptions in employment as well as fish markets. Despite these attempts to bring things under control, bio-economic equilibrium is often eventually reached anyway, and at that point we would expect the lack of profits to drive some fishers out and discourage others from entering the fishery. But as we saw earlier, this does not happen. Long-time producers may have special skills and equipment that are useful only in their fishery, making them reluctant to leave under any circumstances. Fishers who have been residentially stable for many generations may recall better days and stake their future on unrealistic hopes that the fishery will someday return to its former state. The fishing fleets in fisheries suffering from overcapacity tend to consist of old vessels and to be operated by older fishers; neither can be easily relocated or redirected to other economic activities.

Furthermore, though bioeconomic equilibrium is a useful theoretical concept for describing the state of a fishery as a whole, small-scale fishers with their relatively low capital commitments may continue to perceive incentives to increase their production, even after large-scale operators begin to suffer net losses (Tillion 1985; Beddington and Rettig 1984: 3). Less understandable is the thinking that impels newcomers to enter a fishery already suffering from overcapacity. Inexperience, lack of information, the exaggerated claims of fishing-equipment sellers, or a combination of the three, probably account for most of these unwise decisions (Beddington and Rettig 1984: 5).

Maintaining economic productivity and efficiency must be a cornerstone of any management policy, but politicians are reluctant to say much about the virtues of economic efficiency. Perhaps even more than economists, they know that legislating efficiency may work unthinkable human hardships. Understandably, most would rather stress concern for problems that will appeal to their constituencies: alleviating unemployment, raising incomes, and so forth. But there is one objective that most participants in a fishery can usually agree on: achieving some measure of stability in incomes or catch sizes. In practice, however, it is very difficult to ensure stability in most fisheries, since all fish stocks are subject to natural fluctuations to one degree or another. Reducing the extremes of these fluctuations is about the most fisheries managers can hope for, and even this is difficult to achieve if a consistently high level of yield is expected.

Administrative Feasibility and Political Acceptability

Because of the persistent tendency toward depletion in heavily fished, common property, open-access fisheries, some means of regulation is mandatory. Moreover, only governmental action can address proposals to convert these regimes to ones that either grant territorial rights or limit access, or some combination of both. Governmental action is also required to combat such external threats as environmental degradation and encroachment by competing fishing enterprises.

Unfortunately and perhaps unavoidably, political considerations nearly always dominate the policy-making process. Better public awareness of fisheries problems, of alternatives, and of the consequences of various policies can help to enlighten the responsible officials, but even then the decisions will never be entirely free from political influence. Again unfortunately, for many officials the prime consideration is which approach will solidify or erode their political power.

Fisheries policies and regulations must also be acceptable to the people whose behavior they seek to alter or constrain. When fishers cannot understand the regulations, and especially when they feel the regulations are against their best interests, enforcement inevitably produces conflict and can be very expensive. Quite often, it is not the severity of a regulatory regime that angers fishers so much as their perception that it is being inequitably implemented or enforced.

Cost is obviously an important consideration in contemplating a change of regime. Determining what a given strategy will cost in such areas as monitoring, data collection, and enforcement must be as integral a part of the policy-making process as determining its likely efficacy. It may be that the costs of a favored strategy will exceed its benefits.

There is no single best approach, finally, to guide policymakers in establishing management regimes beyond certain principles: finding an appropriate mix of strategies based on their respective costs and benefits; making sure that the affected parties understand the rules and feel they are fair; and not least, systematically reassessing and, if need be, revising whatever policies they settle on.

The Managing of Different Types of Fisheries

Different types of fisheries require different regulatory regimes. The target species, the main gear used to catch them, and the characteristics of the fishers who pursue them all create the need for different approaches. Pelagic fisheries, for instance, are the most important in the world in terms of overall volume, and purse seining is the main fishing method employed. Special considerations in managing these fisheries include distinguishing between short-lived and long-lived species, and the fact that most of the species experience dramatic declines when spawning stocks are heavily fished. For many pelagic species, even after fishing effort is curtailed or halted altogether, the stock may still not recover. Additionally, many pelagic stocks appear to undergo wide natural fluctuations regardless of fishing effort. Finally, because of their open-ocean setting, the regulation of these fisheries requires large expenditures for monitoring, administration, and enforcement.

Pelagic fisheries are mainly the domain of the large, distant-water, industrial purse seiner, although along coastlines where the continental shelf is very close to shore, both large- and small-scale fishers may harvest pelagic resources. In such regions, conflicts may arise between these two very different fishing sectors. Conflicts may also arise within the large-scale industrial sector itself, between those who produce mainly fish meal and target the more abundant younger fish and those who produce fish for human consumption and want to see more mature fish in the stocks. In the absence of regulation, competition between such groups tends to favor the fish-meal producers, since they are able to exploit younger fish before they recruit to the food-fish fishery. The collapse of the North Sea herring fishery is a well-known example of the consequences of this kind of conflict (Beddington and Rettig 1984: 26).

Overcapitalization in pelagic fishing fleets is often an acute problem. Because many species are only available seasonally, it makes the best economical sense to use distant-water fishing vessels capable of making large catches in short time periods. Other economies of scale also favor the larger-capacity purse seiners.

After pelagic fisheries, demersal fisheries are the second-most important in terms of total volume of catch. Demersal fish are bottom-living fish and are usually caught by ground trawling. Like pelagic fishing, this type of fishing is mainly the domain of the large, distant-water, industrial fishing ship. The recruitment of juvenile fish is usually not the critical problem in heavily pressured demersal fisheries that it is in pelagic fisheries.[1] The main conservation problem in these fisheries is *growth overfishing*, or, a level of fishing effort so high that mostly smaller, younger fish are caught. Given more time in which to grow, these fish might have significantly increased the overall biomass of the stock. A related problem is an increase in the proportion of by-catch as the fishery becomes more pressured. Of course, the more by-catch that is thrown back and wasted, the less efficiently the ocean ecosystem is being exploited. Because of the high-seas character of most demersal fisheries, larger vessels are again emphasized, and overcapacity and overcapitalization can quickly become a problem.

[1] There have been few dramatic failures because of recruitment collapse in demersal fisheries. The only notable exceptions are the Arctic cod, gray Pacific cod, and Georges Bank haddock fisheries (Beddington and Rettig 1984: 27).

Fisheries in tropical regions pose particularly complex problems for managers. As Bailey (1988b: 111) notes, "Most tropical fisheries can be characterized as multi-species resources, whose population dynamics are quite complex." The composition of the catch is typically far more diverse than in cold- or temperate-water fisheries, with many different species coming up in a single trawl. Moreover, with prolonged effort the phenomenon known as *ecosystem overfishing* may arise, wherein the fish community is transformed from one dominated by larger species to one composed mainly of smaller species. For fishers, the net effect of this transformation is an overall reduction in the catch, since an increasingly higher proportion of "trash fish," or by-catch, is produced (Beddington and Rettig 1984: 29).

The determination of the most appropriate management strategies for tropical fisheries has been exceedingly difficult. Given the diversity of species in most, regulation by mesh size is seldom useful. The main strategies usually relied on are aimed at keeping catches at relatively low levels in order to avoid distortions in the composition of the fish community. So far, no models having any significant predictive power have been developed to show the impact of various regulatory strategies on tropical fish communities. In heavily pressured tropical fisheries, ecosystem overfishing often prompts considerable conflict between different types of participants. Overall declines in catch rates, for example, as well as the increasing prevalence of trash fish in catches, have been particularly detrimental to small-scale fishers, who tend to produce food for human consumption.

Fisheries with mainly sedentary marine resources, such as clams, oysters, crabs, and seaweeds, or semisedentary or local species, such as lobsters and reef fish, also pose special managerial problems. Here the main problem is usually the relatively great vulnerability of the species to fishing effort. Ensuring recruitment, on the other hand, is usually not a serious concern, since the recruitment of most species is usually little influenced by the size of the adult populations of the stocks. When these fisheries are instituted as common property, open-access resources, they are particularly vulnerable to growth overfishing, meaning that with prolonged effort there will usually be an overall reduction in the larger, more valuable age classes. Properly managed, however, these mainly inshore or near-shore fisheries can be very profitable for coastal fishers.

Fisheries that concentrate on species that are vulnerable at certain times because of their life cycles also pose special problems. Anadromous species such as salmon, for example, are particularly vulnerable to capture during their spawning runs inshore and upriver. Similarly, species that concentrate in large shoals at spawning time—the herrings, for example—are exceedingly vulnerable to capture, as are certain marine mammals during their mating seasons. The main consideration in managing such stocks is ensuring adequate escapement so that recruitment is not totally disrupted.

Tuna fisheries pose a quite different set of problems for fisheries managers (see Clark and Mangel 1977). Because nearly all the tunas are highly migratory, many species move through several national EEZs during the course of their life cycles. Thus the regulation of tuna fishing requires international cooperation. But even then, regulation is difficult, because it is hard to assess the size of these far-ranging stocks; there is usually little evidence of declines in recruitment with declines in stock sizes until a catastrophic point is reached.

MSY, MEY, and OSY

It has been a long time since fisheries managers thought that if their regulatory regimes merely ensured MSY (maximum sustainable yields), their fisheries would remain healthy. The experience in fisheries that were managed in this way has taught us otherwise. We now know that as an open-access fishery develops and its production sector matures, there is a persistent tendency toward overcapacity, resulting in an inefficient waste of capital, even when overall yields are sustained at high levels. In many cases, as competition became acute, the fishery began to lose its most efficient producers—those having the greatest economic and geographic mobility—which in turn only strengthened the forces that tend to move a fishery toward its point of bioeconomic equilibrium. Thus by the time that point was finally reached, these fisheries were being mainly fished by those who were less efficient to begin with. Unfortunately, their relatively high production costs were then passed along to consumers (see C. W. Clark 1976, 1977; Doucet 1984).[2]

As mentioned earlier, the MSY concept is flawed by its assumption that controlling fishing effort on certain fish stocks would keep their numbers and age-class composition in some sort of biological equilibrium. In reality, many marine stocks undergo fluctuations that are quite independent of fishing effort. Eventually, other limitations of the MSY concept were recognized. H. Scott Gordon (1957: 68), for one, was quick to criticize MSY as a realistic objective of fisheries management, contending that it "neglects al-

---

[2] See also Doucet (1984) documenting such a pattern in the Bay of Fundy.

together the fact that in order to catch fish we must give up other valuable goods and services. . . . We even use up other fish . . . to bait the hooks to catch the fish we want." For Waugh (1984: 228), MSY is simply "an inadequate objective because it gives no weight either to the value society places on that yield or to the costs to society of taking that yield."

Among most fisheries managers nowadays, MEY (maximum economic yield) has replaced MSY as the cardinal objective of management. Waugh (1984: 109), one of the leading proponents of MEY as a fundamental objective of management, emphasizes that the concept is theoretically stronger than MSY and capable of being applied to any biological model. Quite unlike MSY, he stresses, MEY is not an equilibrium concept. However, on closer scrutiny, MEY is almost as seriously flawed as the out-of-favor MSY. For one thing, as we have seen, the concept is most useful where there is a stable market for fish—an uncommon situation in most fisheries. And for another, calculating the maximum economic yield requires a precise measurement of the myriad costs and benefits involved in all fishing activity—a formidable methodological and practical problem. Not only this, but in actual practice this purportedly broad and comprehensive concept often turns out to be very limited and narrowly conceived in terms of its ability to prescribe solutions to other key management problems: matters like access rights; the distribution of economic benefits; the role of economic benefits deriving from other, nonfishing enterprises operating in the fishery or of noneconomic benefits accruing to such activities as recreational fishing; the unique or special needs of certain interest groups operating within the fishery; and the fishery's economic link with the rest of the region.

Managing for MEY may also cause undesirable environmental or ecological problems. For example, in the case of high-value fish resources, MEY may require a level of fishing effort considerably above the level of MSY—at a point where the resource is nearly depleted. Other problems may arise, such as excessively high production costs that are passed on to consumers.

Pelagic fisheries in particular may suffer severe ecological problems under MEY. Catch rates may remain high while fish stocks are being seriously depleted. Thus though fishers and managers may feel that their fishery is being soundly managed, collapse can be just around the corner. The catastrophic collapse of Peru's anchovy fishery in 1973 is one of the best-known cases of such an unexpected occurrence. Similar dislocations have been recorded elsewhere, such as in the formerly very profitable herring fisheries in the North Atlantic.[3]

Increasing numbers of fisheries scientists began to stress the need to generalize the MEY concept. Among the first to do so were Milner B. Schaefer and R. Herrington. Schaefer (1957) argued that assuring a people's local food supply is a fundamental concern that should be incorporated into the concept, while Herrington (1962) stressed the need to incorporate the benefits of increased employment. M. James (1959), another early critic, pointed out that MEY does not adequately incorporate political, social, and cultural considerations. In another vein, R. Taylor (1963) suggested that a measure of economic inefficiency and overemployment was justified in South Australia's whiting fishery in the interest of the more urgent objective of decentralization. Others have argued for the integration of social benefits accruing from fishing activity that are not usually incorporated into the conception of MEY (Alverson and Paulik 1973) and for recreational fishing as an essential consideration (E. N. Anderson, Jr. 1975; Radovich 1975).

Economists remain the most stalwart defenders of MEY as it is usually conceived, and many strongly object to the various proposals for generalizing the concept mentioned above. Harold C. Frick (1957: 685) spoke for many of them in his opposition to the notion of maximizing employment at the expense of overall economic efficiency: "Providing a maximum opportunity for employment is not a valid objective of management policy, if it is to be accepted that social welfare is best served by facilitating the flow of productive resources, including labour, into the most (value) productive areas."

The opinions summarized above stand on opposite sides of a great philosophical and theoretical divide, presenting policymakers with dilemmas that are not easily resolved. On one side is the traditional school, consisting mainly of orthodox economists who emphasize that extracting the maximum economic rent from the fisheries as efficiently as possible ultimately provides the most good for the most people. On the other side is a growing and diverse group, predominantly social scientists (other than economists), who stress that pursuing the objective of economic optimality to the extreme works unthinkable hardships on human beings and also

---

[3] The collapse of the North Atlantic herring fishery, as well as other notable collapses in pelagic fisheries, is discussed in Beddington and Rettig (1984); Butterworth (1983); Saetersdal (1980); Troadec et al. (1980); and Ulltang (1975, 1980).

poses great risks, should anything upset a high-strung management regime that approximates MEY.

This latter group, nearly all of whom are fishers' advocates of one sort or another, echo E. F. Schumacher (1973) in wanting to see "economics as if people mattered." They have also been the main voices calling for further clarification of the objective of managing the fisheries for optimum social yield, or OSY. In theory, as we have seen, this concept would see biological, economic, social, political, and other variables perceived as desirable by human beings incorporated into a single objective function for managing a fishery. However, because so far OSY remains little more than an ideal, both in conception and in application, it is easily assailed by its critics. Many concede that it is a worthy goal but point to its greatest weakness: the difficulty of making it work. Waugh (1984: 113) calls optimum social yield "an omnibus objective, meaning all things to all men." In his view, "The composite function becomes so vague that it loses any operational significance" (p. 111). Waugh points particularly to the problem of operationalizing diverse values. He agrees that the value of recreational fishing, for example, constitutes an integral part of a fishery's overall economic yield, but wonders how that value can be economically compared with values in the commercial sector, since most recreational fishers "see value in the activity and not just the catch" (p. 112). Here, it seems, he is unintentionally reinforcing the position of proponents of the OSY concept: that human beings, whether they fish for fun or for their livelihood, obtain important other values from this activity that can be conceived of separately from its strictly economic import.

Still, theorists who stress the efficacy of MEY as a cardinal objective of fisheries management are unlikely to relinquish that position easily to the advocates of OSY. Some feel that the advocates of OSY have not understood how broad the scope of the MEY objective actually is. Others charge them with not understanding the large body of sound economic theory that supports it. Unfortunately, many of these partisans too easily dismiss the importance of certain other fisheries management problems—the promulgation of regulations that are acceptable to fishers, for example—seeing these as exogenous to the problem of managing a fishery for MEY. Such a stance arbitrarily and unjustifiably partitions the management problem. Formulating fisheries management regimes that are acceptable to fishers is no mere exogenous consideration. It is a prime consideration in fisheries management and will remain so until human behavior and human-value considerations are better incorporated into the formalistic frameworks now being used in fisheries management.

Theorizing about how to optimize biological variables such as the quantity, size, and age cohorts of fish seems simple compared with theorizing about optimizing human variables such as the quality of life and the standard of living. Nevertheless, if OSY is ever to be integrated with the well-developed theoretical conceptions of MSY and MEY, it must somehow be conceptualized into a more formal calculus that incorporates both quantitative and qualitative variables.

Daunting as the task is, it need not bring us to a methodological frontier that we cannot cross for lack of guidance and experience. The now fairly well-developed methodological approaches of social-impact assessment, for example, have much to offer. Moreover, the formal frameworks of finite mathematics, particularly those known as games of strategy, linear programming, and directed graphs are capable of incorporating qualitative as well as quantitative values into single analytical frameworks having quantitative salience. These have already enjoyed great success in business applications, and more recently they have made important contributions to the development of method and theory in the social sciences.[4]

In 1986, for example, Ira Buchler, M. Fischer, and I published a linear-programming analysis of economic decision making in a primitive society that had no medium of exchange or commercial markets. Local "societal" values, such as maximizing the community's security by ensuring the production of a minimal level of subsistence food supplies and maintaining the overall work effort at customary levels, were incorporated into a single objective function, along with other, more purely economic values, such as obtaining maximum agricultural production from various types of land, each of which was associated with different levels of productivity, labor input, and risk. In the end, the ethnographic data showed that the local

---

[4] On formal mathematical models in anthropology and related social sciences, the reader might start with the following literature. For the origins and philosophical and theoretical underpinnings of game theory, as well as its potential applicability in the study of human economic behavior, Von Neumann and Morgenstern (1944). For an easy-to-read primer on game theory, Williams (1954). For finite-mathematical frameworks and their potential applicability in the social sciences, Kemeney et al. (1974: esp. ch. 8). For game theory and linear programming in anthropology, Buchler and Nutini, eds. (1969); Buchler, Fischer, and McGoodwin (1986). For directed-graph theory in anthropology, Hage and Harary (1983). For mathematically sophisticated analyses in general in social and cultural anthropology, White (1973). And for analyses relying on modeling paradigms other than the canonical forms characteristic of finite mathematical frameworks, Ballonoff (1974).

populace was doing exactly what the model suggested it should be doing for optimum agricultural production. In other words, by integrating qualitative human values with quantitative economic values, we demonstrated that the local populace was behaving "rationally" by modern standards, as well as by their own.

Others have tackled the problem from another angle. Gerald D. Brewer (1986), for instance, urges that different types of ecosystem models, simulations, and formalized games—what he terms "policy exercises"—be incorporated into the policy-making process for managing natural resources.

Several students of the fisheries (e.g., Berkes and Kence n.d.) have observed that fisheries managers and users are in essence locked into a conflict situation analogous to the formal, mathematical game known as "the prisoner's dilemma," in which "players" can only optimize their potential gains and minimize their potential losses by anticipating what their opponents may do. Thus, as Fricke (1988: 7) sees it, "The 'prisoner's dilemma' game provides a general framework of analysis for the evolution of mutual cooperation."

The development of still more sophisticated analytical tools may reveal that many of the behaviors of fishers that are today regarded as "irrational," or counterproductive to effective management, are actually quite rational indeed. In a larger sense, these formal sociological models may help managers better understand why fishers act as they do, something that may in turn contribute to the development of more appropriate management policies. Policy analysts, scientists, and fisheries managers should therefore redouble their efforts to bring about a more formalized and comprehensive conception of OSY. Just because the problem is exceedingly difficult is no reason to set it aside.

Development

Finally we come to a crucial question for policymakers—fishery development. It is essential that management never be considered apart from this matter, for actions taken to promote one may negatively impact the other. The best way to avoid this problem, perhaps, is simply to integrate the two concerns under one administrative framework, domestic or international (FAO 1983: 24).

A fishery can be developed at an astonishing rate, thanks to the relatively great mobility of fishers and fishing enterprises, sometimes far outpacing scientists' ability to monitor the effect on resources. Hence managers are often ill-equipped to respond in a timely manner to implement revised and more appropriate managerial regimes. Moreover, resources may seem so plentiful and profits so high that it is difficult to conceive of the possibility of an impending catastrophe or utter collapse. Many case studies are available documenting the serious dislocations and disasters that have occurred in fisheries that were developed too rapidly; Peru's anchovy fishery is probably the best known (Glantz and Thompson, eds. 1981).

Over the past four decades, many of the world's developing nations have sought to exploit their fisheries and have received assistance from development agencies, development banks, and so forth. In many cases, such assistance has led to healthier and more productive fisheries, but just as often it has led to new and serious problems. Clearly, then, policymakers contemplating action in this domain should always integrate both management and development concerns (Beddington and Rettig 1984: 2). I will return to this point in the final chapter.

TEN

Strategies and Dilemmas

A fishery's overall management policy and management regime—its grand strategy, if you will—usually brings into play a mixture of disparate management strategies, each of which presents managers with an array of choices, or substrategies, on how they will be implemented. But every grand strategy ought to have one common feature: some mechanism that gives managers the flexibility to juggle the emphasis given to one strategy or another and to add or drop a strategy as changing conditions require.

It is not only human activities and considerations that prompt the managerial dilemmas so often associated with most strategies. The unpredictable sea itself, with its marine populations fluctuating in response to dynamics that are still often shrouded in mystery, frequently confounds managerial policies and regimes as much as a fishery's human participants. This inherent unpredictability is still underestimated by some fisheries managers, partly because of the short time horizons with which they often have to work and partly because they often do not have good data on marine resource levels.

By taking a long-term view of a fishery, we discover that fishers are also often subject to fluctuations. Initially, a few small-scale fishers enter a fishery, let us say. Then, as the years pass, some prosper enough to become the employers of other formerly independent producers. Simultaneously, vessel sizes increase and capture technologies become more effective. Then, sometimes abruptly, the bioeconomic equilibrium point is reached, and a large number of fishers suddenly leave the fishery. Thus the fishers in a particular fishery—much like the fish they seek—may comprise certain age classes or cohorts who are at the same stage or stages of development.

Fisheries managers, then, must understand not only the complex dynamics of marine populations, but also the population dynamics of the fishers they seek to manage, appreciating that both undergo changes that are sometimes hard to anticipate (see FAO 1983: 11–12). An additional perplexing problem, in this connection, is the geographic mobility of fishers. As Acheson (1988b: 63) notes, changing fisheries is one of the most common adaptive strategies fishers employ. "This," he states, "means that regulations to reduce effort on a given species often means nothing more than shifting that effort to other fisheries which can cause problems in those fisheries in the future."

At the most basic level, the process of fisheries management involves solving two fundamental problems. The first is conservation: deciding what amount of fish can be harvested on a sustainable basis. The second is allocation: deciding who benefits, in what ways, and to what extent (C. L. Smith 1988). Currently, seven basic strategies are employed to satisfy both concerns. They are:

1. Closing areas
2. Closing seasons
3. Restricting gear and technology
4. Establishing aggregate quotas on total allowable catches (TACs)
5. Attempting to stimulate fisheries growth or to control fishing effort through monetary measures such as subsidies and taxes
6. Limiting entry
7. Instituting various forms of private property rights

Strategies 1–5 were developed mainly in common property, open-access fisheries where marine resources were already being pressured to their limits, most often by industrial fishers. The other two—limiting entry and establishing property rights—are ancient strategies now enjoying revived interest, particularly with the new trend away from managing fisheries as common property, open-access resources. Some of these strategies are appropriate for the management of small-scale fishers; others are not.

The list does not include various proposed new strategies, some of which are now being tried out. These include fishers' associations that manage fisheries with little influence from governmental entities and "co-management" regimes in which fishers share joint managerial authority with government officials. I will discuss these strategies in some detail in the final chapter.

In what follows I examine each of the seven main strategies now employed, with special attention to both its costs and benefits and its drawbacks from a managerial point of view.[1]

Closing Areas or Seasons

Indirect methods for reducing fishing effort like closing areas or seasons and placing restrictions on fishing gear were among the first strategies to be adopted in modern fisheries management. Such methods are fairly easy to implement and enforce because they are straightforward and perceived as applying more or less equally to all producers in a fishery.

Closed areas or seasons are generally imposed when fish stocks are down or in poor condition, or during down markets, or when poisonous species might be caught, or when nursery grounds and juvenile stocks are in need of protection. Areas are sometimes put off-limits for other, essentially political reasons: to reduce conflicts, in the interests of socioeconomic equity, and because of national policy concerns. Closures are also sometimes used for safety reasons during seasons of unstable and bad weather (FAO 1983: 7).

The main objective is usually conservation. Geographical areas designated as conservation districts may be established to protect a portion of a marine stock from exploitation or to allow it to recover from overexploitation. Protected areas are also often established where large aggregations of spawning stocks or juveniles are known to congregate to permit adequate escapement.

Many developing countries have recently established "coastal belts" to protect their small-scale fishers from competition by larger-scale fishers. To be effective, this policy requires the collection of detailed information on the size, composition, and age structure of the fish stocks present in the coastal belts. Such data-gathering may be prohibitively expensive for a developing nation. This approach to management has other drawbacks as well. It is not particularly useful when the main stock is highly migratory and can be exploited by industrial fishers before the fish move inshore. And if coastal belts are to achieve their purpose, a fairly high degree of surveillance and enforcement is required; this too may be prohibitively expensive for a developing nation.

---

[1] This discussion relies mainly on Beddington and Rettig (1984), Bell (1978), FAO (1983), and Waugh (1984), who are cited throughout and whose insights and experience I gratefully acknowledge.

The main advantages of this strategy are twofold: the policy is easy to implement and it is extremely flexible. Both areas and seasons can be opened and closed fairly quickly, as updated information about the stocks suggests. However, the main disadvantage is a serious one indeed; when employed exclusively, this strategy does nothing to discourage the development of overcapacity.

Two types of closed-season regimes may be established. The first is relatively simple and merely entails closing a fishery during certain times of the year—when juvenile or spawning members of the stock would be too easily caught, for instance. The second is considerably more complicated. It entails closing a fishery when catch size or the catch composition per unit of effort falls below some desired level. Implementing this second type of regime requires close and continual monitoring of catches and is therefore expensive. Also, it is only useful when there is a strong relation between overall catch or catch composition and fishing effort (Beddington and Rettig 1984: 11).

Closed seasons are a great way to protect vulnerable stocks, but the strategy falls sadly short otherwise. In some sense it might even be said to encourage overcapacity, for fishers will strive to maximize their production during the open season by investing in more expensive and effective fishing vessels and fishing technology or trying by other means to extract as much as they can from the fishery. The inevitable result is increasingly short fishing seasons and disrupted fish markets because of severe discontinuities in supply. There is also sometimes a fall-off in the quality of the product reaching consumers because of the intensity with which the fish are caught and the haste with which they are brought to market (Beddington and Rettig 1984: 12).

In sum, closed areas and seasons are useful strategies for meeting conservationist objectives, but are most effective when used in conjunction with other strategies that resist the development of overcapacity. In their simpler forms, they are fairly easy to administer and enforce. This is especially true of closed seasons; closed areas, on the other hand, particularly in fisheries exploited mainly by small-scale fishers, may be difficult to enforce if the fishers do not have charts or navigational gear to tell them where the boundaries are. Moreover, when areas are closed in the interest of protecting an overexploited or vulnerable species, fishing for other species that are neither overexploited nor particularly vulnerable is shut off (FAO 1983: 8).

Restricting Gear and Technology

Fishing gear and technology are typically restricted by setting maximum sizes on net meshes, particularly in trawls and gill nets, and fishhooks; requiring escape gaps in fish traps or pots, particularly in crab and lobster fishing; and limiting vessels to a certain size or horsepower.

Despite the ubiquity of this strategy in modern fisheries management, it has been surrounded by controversy throughout its long history. The strategy appeals to most managers because it is easy to explain to fishers, easy to implement and enforce, and usually effective in accomplishing its intended biological objective—generally to see that only the larger fish are caught. For fishers, however, the strategy can mean retiring still-useful equipment and purchasing expensive new gear. Many, particularly small-scale fishers, dislike it simply because it limits their freedom to fish in whatever way they wish, using the technologies they prefer. Certain technological restrictions have also been criticized by economists and policy analysts, as well as by fishers, on the grounds that they constitute mandated inefficiency.

When technological restrictions can be shown to lead to higher quality and more valuable catches, fishers usually greet them with enthusiasm and comply with little coaxing. Indeed, in some places fishers have voluntarily restricted themselves to certain types of gear in order to maximize the quality and value of their catches. However, where fisheries are already overfished and economically depressed, fishers usually resist the imposition of gear restrictions and will attempt to evade them.

The strategy is most appropriate where it is important to ensure a certain age of first capture by permitting juvenile stocks to mature and become more marketable. It is also best employed in fisheries characterized by high rates of gear replacement or in fisheries undergoing rapid rates of technological change and innovation. It is far easier to require that fishers use certain types of gear or fishing technologies where they must regularly replace or upgrade their fishing gear than where the main gear or technologies they employ have long useful lives. Finally, this strategy has the virtue of helping to prevent the development of overcapacity because it can attack the problem directly.

Mesh regulations work particularly well in single-species fisheries or in fisheries where only a few species of similar size are targeted for capture—in the North Atlantic and North Sea fisheries, for example. They are of little use in multispecies fisheries, particularly those in tropical waters, where trawlers typically catch a large number of species, each with its own optimum mesh size (see Beddington and Rettig 1984: 13; FAO 1983: 8). Mesh regulations are also a useful strategy for promoting the interests of a certain class of fishers. Banning small-mesh nets, for example, may benefit small-scale fishers who fish for human consumption and impede industrial fishers interested mainly in producing fish oil and fish meal.

Fishers sometimes oppose the enactment of mesh regulations because catch rates inevitably fall when they first go into effect. This is likely to be a serious problem unless the targeted species are short-lived and fast-growing. Furthermore, while the benefits of mesh regulations are easy to show in theory, it may be several fishing seasons before they can be discerned. Natural variations in the recruitment of different age classes in a fish species may override the effects of new mesh regulations for a time, masking the fact that they are having their intended effect. Fishers who were initially persuaded by the benefits of complying with the new regulations may then become skeptical and do their best to circumvent them, something that can only confound the management regime and worsen working relationships between them and the fishery's managers (Beddington and Rettig 1984: 12).

Establishing Aggregate Quotas (TACs)

Regulating a fishery by establishing a total allowable catch (TAC) for a fishing season is a very complicated and involved process. First, based on biological statistics derived mainly from sampling and from the monitoring of catches, scientists must estimate the total catch a fishery can sustain in future fishing seasons. Then working groups and committees of scientists, fisheries managers, policymakers, fishers, and other interested parties must negotiate and eventually agree on the TAC for the forthcoming season. In this stage, political and economic considerations are often more influential in determining the TAC than biological ones. Next, crucial policy decisions must still be made on how the TAC is to be divided among the various producers, and once this is accomplished, various regulations have to be established defining permissible means of access, fishing times, methods or gear, and so forth. Finally, enforcement systems and systems for monitoring production must be established, so that the regulatory authority knows when the TAC is reached and it is time to call a halt to fishing. The TAC strategy works best in high-seas fisheries, especially those that are exploited by industrialized fishers from several nations. It also works best in single-species fisheries. Calculating the TAC for multispecies fish-

limitation
of TAC

eries like those in the tropics is usually too complex to have much practical value. Moreover, because the main aim of establishing a TAC is biological conservation, the strategy by itself will not prevent the development of overcapacity.

Management emphasizing TACs requires considerable scientific and technical manpower. In the words of John F. Caddy (1984), these regimes are "information hungry." Therefore, the TAC strategy is not appropriate for fisheries of poor nations, nor is it likely to be cost-effective in ocean fisheries exploited by a group of poor nations.

Determining a TAC is analogous to determining a fishery's maximum sustained yield (MSY). Thanks to mathematical formulas provided over thirty years ago (Beverton and Holt 1957), the calculation itself is a rather simple matter. The knotty problem is obtaining an accurate assessment of the size and biological structure of the targeted stocks. Various techniques are used to this end. Though more costly, direct data-gathering methods are more accurate than indirect ones. These methods rely on random samples of catches brought on board research ships, acoustic (sonar) surveys, and, increasingly, remote sensing by aircraft and satellites. Using catch-monitoring operations as a source of data is perhaps the commonest indirect method. Others include relying on dynamic models like the Virtual Population Analysis model or on models that estimate the stock's future size and age structure based on catch data (Beddington and Rettig 1984: 8).

Whatever the techniques used, ensuring the accuracy of the data is problematical. Even direct methods sometimes yield ambiguous or inaccurate information. Acoustic surveys conducted over shallow shoals populated by several different species, for example, will yield data that are nearly impossible to interpret. And figures derived from short data-collection histories are likely to be unreliable. Furthermore, the quality of the data on catches tends to deteriorate over time as fishers find more ingenious ways to underreport them.

Other problems also arise when fishers do not conform to the regulatory regime in good faith. For instance, in some fisheries the TAC is defined as the point at which the by-catch reaches a certain proportion of the catch. At that point, fishing for the targeted species should cease, but fishers may circumvent this regulation by throwing away the by-catch before it can be measured. Or when fishers' quotas are expressed in terms of a total allowable catch for a particular species, they may throw away the less valuable smaller fish of that species so they can continue fishing, thus wasting large quantities of the targeted species.

Once a fishery's TAC is determined, there remains the knotty equity problem of how to allocate it among the various producers. Experience has shown that producers will attempt to take all the stock they can before a TAC is reached and the fishery closed, using ever-larger and more powerful vessels to get to the fishing grounds as quickly as possible so as to maximize their production. In fisheries that have become overcapitalized in this way, overcapacity is a nearly inevitable long-term consequence. The TAC will be reached in ever-shorter periods of time, causing severe dislocations in markets and in fishers' employment patterns—a brief feast followed by a long famine. TAC regimes must therefore be augmented with other provisions that attempt to mitigate these problems. Otherwise, they can result in successful management from a biological perspective while promoting social and economic catastrophes. In international fisheries, the allocation problem is compounded by the need for agreement among several nations. This usually involves delicate and complicated negotiations, in which both historical concerns and each nation's foreign policy are important considerations.

Whether a nation has received a proportion of a TAC through international negotiations or has established a TAC in a fishery of its own, the subsequent allocation of the TAC among the producers presents an excellent opportunity for addressing social, economic, and political concerns. Allocations can be used to promote certain sectors within the fishery or in the society at large over those that have become inefficient, obsolete, or problematic from some other perspective; to increase social equity; to counter tendencies toward overcapacity by favoring fishers who use only certain types of vessels or fishing gear; and so on. Allocations designed to further the interests of small-scale fishers are more feasible in developed nations than in poorer ones because of the expense of the monitoring effort.

Because compliance is key to the effectiveness of TAC regimes, enforcement tends to be one of the most costly aspects of these regimes. Though both aircraft and seagoing vessels may be used to this end, managers have relied chiefly on on-board observers up till now. This is still a costly enforcement tool, and one that is often not feasible on smaller boats with limited crew space. It also grates on many fishers, who see it as an infringement of their freedom; from their point of view, the practice amounts to having a spy or a policeman on board at all times.

Though fishers are the group most likely to be held liable for violating TAC regulations, there is a growing mood to extend the enforcement effort to the point of first sale, charging buyers and processors who knowingly accept production that is in excess of the quota or inaccurately reported with receiving stolen property (Beddington and Rettig 1984: 11).

Apart from providing data—either by complex reporting procedures or by being boarded and sampled—fishers have little involvement in most TAC regimes. But this seems likely to change, for there seems to be an increasing recognition on the part of fisheries officials that if fishers understood the regulations and were convinced that they were both necessary and fair, they themselves could become the best agents for enforcing them. Clearly, then, when fishers assert that certain regulations are so difficult to comply with that they are practically forced into breaking the law, fisheries managers ought to consider changing those particular regulations rather than continuing a futile enforcement effort (see FAO 1983: 5). Engaging fishers in the process might thus work a considerable cost savings. Monitoring and surveillance costs could be much reduced if the tasks fell to fishing boats already operating in the fishery instead of aircraft or vessels operated by the regulatory agent. And fishers could do the job of logging the amount and composition of catches far less expensively than employees of the regulatory agent, assuming they could be persuaded to comply with the regime in good faith.

Using Monetary Measures

This strategy comprises several substrategies; the most common are subsidies, tax incentives, direct taxes, taxes collected on the basis of gross weights of landings, and license fees. The collection of license fees, while an important monetary means for controlling fishing, is examined in the next section as a main substrategy for limiting entry.

Because of overriding political considerations, governments and politicians are usually reluctant to impose new taxes but are often enthusiastic about legislating subsidies as a means for stimulating development in the fisheries. The great appeal of subsidies has often caused them to become overused, creating more problems than they resolved. In fisheries suffering problems associated with overcapacity, for instance, a wise course would be to withdraw subsidies and increase taxation on those sectors most responsible for the problem. It is not a course, however, that many politicians would be willing to stake their careers on.

A related problem is the lack of decisiveness or narrowness with which monetary measures have often been implemented, with politicians again not wanting to seem to go too far by overtly favoring one group over another. Unfortunately, when monetary measures are employed halfheartedly and indecisively, they may only muddle and complicate the task of fisheries management. So, while in theory such measures hold promise as a useful strategy for fisheries management, in practice the political implications have often merely confounded and frustrated it.

Governments often attempt to improve the health of their fisheries, particularly their developing ones, through subsidies and other financial support programs. A whole array of direct subsidies has been attempted here and there: the subsidizing of fuel costs, the provision of low-interest loans for gear acquisition and vessel construction, the support of fish prices, generally by restricting fish imports, and others. In fisheries suffering from overcapacity, direct subsidization has often taken the form of scrapping or buy-out programs that reimburse fishers who are willing to leave the fishery. Many developed nations even have funds for paying fines when their fishers are charged with violating the fisheries of other nations.[2]

Indirect and less-focused means of subsidization have also figured importantly in the attempt to stimulate the health of fisheries. Examples are the sponsorship of research and development for technical or other improvements; public expenditures for fish stocking; dredging projects and other improvements of the marine environment; and the constructing of new port facilities. Some indirect means that have been used are more human-oriented, such as regional programs to develop alternative occupational opportunities in fishing communities or to provide social services to their inhabitants (Beddington and Rettig 1984: 21).

But too often the intended benefits of subsidy programs have negative consequences. Many developing countries, for instance, find it hard to resist the temptation to seek assistance from international development agencies, banks, and others for fisheries improvements. When taken indiscriminately and without adequate consideration of future consequences, such assistance can lead to the adoption of inappropriate fishing technologies and hasten the development of overcapacity, resource depletion, and social inequity. Once these trends start, they are all but impossible to reverse. At that point, governments can find themselves under considerable pressure to initiate or install direct subsidies, in effect encouraging overdevelopment and aggravating an already dire situation (Beddington and Rettig 1984: 21). The alternative, to withdraw subsidy programs in un-

[2] The purpose in most cases is to protect fishers from undue victimization as pawns in international disputes. U.S. fishers are protected under the Fisherman's Protective Act of 1954.

healthy fisheries, particularly those that have become overcapitalized, inevitably creates additional economic strains and considerable ill-will, and can have severe political implications. In essence, it amounts to leaving fishers to shift for themselves just when they are hurting the worst.

Subsidy programs can bring about good results when they are employed to correct imbalances in a nation's distribution of economic benefits: providing loans to disadvantaged small-scale fishers who lack access to credit, for example. Such programs make the most sense when they are designed to stimulate development and then are gradually withdrawn as the fisheries mature, before the subsidies have brought about excess capacity. Thus, in considering the use of various types of subsidies, policymakers should also consider whether some can be shut off more easily than others. Discontinuing low-interest loan programs, for example, may be more easily accomplished than withdrawing subsidies for fuel and gear or price-support programs.

Besides promoting overcapacity, the injudicious use of subsidies can prompt other problems. Fuel subsidies, for instance, may encourage the use of more expensive and less efficient means of propulsion,[3] or benefit the industrial fishers at the expense of the more fuel-efficient small-scale ones, or permit fishers to range more widely and harvest stocks that were previously protected by their inaccessibility (FAO 1983: 17–18).

After subsidies, taxation has been the most favored monetary mechanism for improving the health of fisheries. Like subsidies, taxes are generally levied in pursuit of broad policy objectives, not for the purpose of regulating fishing effort as such. Fishers are taxed, both directly and indirectly, in a variety of ways. They may be required to make payments based on the size or value of the fish they land or to purchase licenses for their vessels and gear. They may be charged berthing fees in public port facilities. Some must make contributions to support the coast guard's rescue capabilities; others must pay for vocational and safety-training programs or purchase permits before they are allowed to work in a fishery. In fisheries managed under TAC regimes, fishers are often required to provide food to the regulatory agents posted on their vessels. Business income

[3] This was the case in the fisheries of St. Lucia, West Indies, I studied in 1984. Because the marketing structure favored fishers who were able to make a quick and early return to their communities with their catches, fishers could gain a decisive competitive edge by using fast and fuel-hungry gasoline-powered outboard engines, with the result that the more fuel-efficient but slower diesel engine was hardly to be seen. The government's subsidization of fuel costs, which did not distinguish between gasoline and diesel fuels, only aggravated the problem.

taxes may also have to be paid. And in developing nations in particular, fishers may be required to contribute their labor for habitat or infrastructural maintenance and improvement.

To the extent that taxes reduce fishers' net incomes, they may serve to curb fishing effort. But fisheries managers rarely consider making this part of their grand strategy. For a start, taxation programs of any kind are rarely popular and are politically difficult to initiate, though fishers may enthusiastically support new taxes when they perceive direct benefits, such as programs for improving aquatic habitats or establishing and maintaining fish hatcheries that produce brood stock. Most taxes are in any case imposed and collected by agencies whose principal concerns lie elsewhere. Moreover, governments tend to levy taxes on the basis of people's ability to pay, not on their use of public resources.

As several fisheries analysts, James A. Crutchfield (1979) for example, have cautioned, it would be dangerous to rely on taxation as a main strategy for controlling fishing effort over the short term. Experience has shown that tax regimes are usually unable to respond quickly to changing conditions in a fishery. Time is eaten up by all the activities needed to get a tax program in place: the collection and careful analysis of data by scientists and fisheries managers, and then the additional effort of convincing legislators that a change is necessary. And there will be still further delays, assuming a change is enacted, before the new regime will significantly alter fishing effort, because it may be some time before fishers switch to other modes of fishing or other fisheries, or leave the fishery altogether. For this reason, the strategy of taxation is best used for dealing with problems on a long-term basis; taxes could be levied on chronically overfished stocks, for example, and plentiful stocks made tax-free (Beddington and Rettig 1984: 19–20).

As things stand, implementing tax regimes in fisheries to defray the costs of regulation, enforcement, and environmental protection, however ideal in principle, is unfeasible. In most nations tax revenues collected from the fisheries sector fall short of public expenditures made for its benefit. Moreover, enforcement can sometimes be prohibitively expensive, particularly where fishers and other participants go to great lengths to avoid paying their taxes. In those situations it is often wiser simply to drop the tax regime and explore other means of raising revenues. Finally, placing taxes on fishers and fishing effort is usually not urgently needed, in the way that the withdrawal of subsidies can be. And as we have seen, however promising tax programs appear to be in theory, they are of little or no help in dealing with problems requiring fast action. As a long-term preventative against the development of overcapacity, however, tax regimes can be very effective.

One last word about subsidies. From a national perspective, to make the general public subsidize fisheries without sharing in the benefits is patently unfair. Thus subsidies are best used not to assist a single group—fishers—since this can quickly and directly lead to overcapacity, but to pay for the kind of improvements in the infrastructure of a fishery that would make life better for fishers and nonfishers alike.

Limiting Entry

Probably the single-most effective strategy for preventing or redressing the problem of overcapacity and for bringing fishing effort under direct control is to limit access to the fishery. It is also a strategy that has been increasingly proposed as a reasonable alternative to more traditional means, such as quota systems and gear restrictions, programs that have often failed to prevent the collapse of fisheries and are costly to implement.

As one would expect, most such proposals are vigorously opposed by fishers, and attempts to implement the policy have met stiff resistance, especially in highly developed fisheries.[4] In fisheries already nearing overcapacity, the mere threat of limiting entry may even prompt fishers to engage in greater excesses of fishing effort. Thus closed areas and seasons, gear restrictions, and TACs, even when of dubious efficacy, tend to be more politically acceptable because they seem to apply to everyone equally, discriminating against no one (Bell 1978: 153).

Unfortunately, political considerations may overwhelm equity considerations when the time comes to decide the crucial questions of who will qualify for entry and who will not. Some of the methods relied on to settle the question are selling or auctioning licenses to the highest bidders, giving precedence to the producers who have historically worked the fishery, and favoring full-time over part-time fishers or small-scale over large-scale fishers. Regional and national socioeconomic policy concerns such as unemployment may also play a role (FAO 1983: 14–15). Furthermore, it will

---

[4] Petterson (1983) describes the extreme difficulties encountered in the attempt to limit entry to the Native American fisheries around Bristol Bay, Alaska. He shows how the objectives of Alaska's Limited Entry Act of 1973 were frustrated by both the regulatory policy and the implementational process itself.

usually be necessary to decide whether to limit the entry of vessels or only the entry of fishers. In this connection, a sentiment expressed by a Gloucester fisherman seems pertinent: "Boats don't fish, people do" (M. L. Miller and J. Van Maanen 1979).

There are three substrategies for limiting entry: first, imposing various types of licensing requirements that limit the number of fishing vessels or fishers or that limit vessels to certain sizes or power plants, or otherwise restrict the type of fishing gear that can be employed; second, allocating— perhaps by auction—transferable quotas to individual fishers or fishing enterprises; and third, limiting entry through monetary means, such as the imposition of taxes.

Licensing programs are obviously best undertaken while a fishery is still developing, avoiding the problem of forcing out existing participants. Though previously unfettered fishers in already developed fisheries are sure to oppose any restrictions, they might be more disposed to accept them if they can be made to see that the stocks have dangerously declined. This is best achieved through educational efforts backed by sound data and explanations of fisheries principles that fishers can readily understand.

Limited licensing programs are often regarded as useful supplements in TAC regimes. These are appropriate primarily where fishers can underwrite some of the high costs entailed and are of little use in fisheries worked mainly by small-scale fishers.

Licensing regimes that merely limit entry to a certain number of vessels or fishers cannot by themselves prevent the development of overcapacity and depletion, since increases in the size and effectiveness of vessels and gear or intensified effort by fishers would still move the fishery toward depletion. Consequently, additional stipulations covering these possibilities must be part of any licensing scheme. So, though such schemes initially seem to offer a simple and straightforward management tool, they often turn out to be very complicated and costly (Waugh 1984: 139; FAO 1983: 22).

In contemplating a licensing scheme, policymakers must confront three questions: how many licenses to grant; who will get them; and whether the licenses will be transferable. As Waugh (1984: 135) observes, "these three questions involve considerably more than economics and are often exercises in politics, industrial relations and law." In general, licenses tend to be issued to vessels rather than to individual fishers.

The first issue cannot be resolved until fishery authorities have the fullest possible data on the natural stock and fishing fleet of a fishery, and on the cultural characteristics of those who work it. The first task is to make an estimate of the MSY of the main stocks. Then, on the basis of the catch rates of the vessels currently operating in the fishery, the total allowable number of vessels can be determined. It is wise to estimate the total number of allowable vessels conservatively, erring on the low side, since the productivity and efficiency of the fishing fleet will often increase in the future, especially once the regime takes hold and severe depletion is avoided. Of course, these considerations become much more complicated when a fishery is exploited by diverse types of fishers, by vessels having widely different productive capacities, and so forth (Beddington and Rettig 1984: 13–14).

In summary, though licensing programs, which have been the main substrategy used for limiting entry into fisheries, are surfacely appealing as an easy way to address fisheries problems, they are no simpler than most of the other, more traditional strategies commonly employed. The implementation of licensing programs requires much prior research and data gathering, and then considerable collaboration, mediation, and negotiation among the various participants in a fishery. Moreover, such regimes are usually rather costly to administer and enforce.

The other two substrategies for limiting entry, quota systems and monetary measures, are used less often. Generally, in a quota system, the MSY is first determined and then divided into parcels that are sold to the highest bidders. Many fisheries managers find this substrategy attractive because it offers the advantage of great flexibility and can also provide revenues to defray management and enforcement costs. In actual practice however, it is usually confounded by two nearly insurmountable problems. First, like all quota schemes, it is expensive to administer and difficult to enforce; and second, it is often vigorously opposed by less-affluent fishers as unfairly favoring those who can push the bid price up (Waugh 1984: 138–40).

Limiting entry by taxation is favored in some circles as a way of simultaneously helping governments to receive economic benefits from their fisheries and bringing fishing effort under control. Taxes are usually levied on landings or other measurable units of fishing effort. However, as we saw earlier, in the section on monetary measures, this is an ill-advised strategy if the purpose is to control fishing effort, because tax regimes are characteristically unable to respond quickly to changing conditions in a fishery.

Establishing Property Rights

To many contemporary fisheries managers who are accustomed to thinking of fisheries as typically common property, open-access resources, regulation through a system of property rights may seem exotic. But, as described in an earlier chapter, this is a very ancient practice and is still found in many parts of the world. It is also a strategy that increasing numbers of specialists regard as holding great promise for the regulation of small-scale fishers, particularly in the developing countries whose large, diverse, and impoverished populations of fishers are difficult to manage by the more conventional, and usually costlier, means employed elsewhere.

The main appeal of instituting a system of property rights is that fishers who consider certain fish stocks or fishing grounds their own property may voluntarily restrain their fishing effort and develop greater concerns for conservation and management. In some cases, in fact, fishers enjoying property rights have recommended restrictions on fishing effort that were even tighter than those recommended by external agents (FAO 1983: 18). Such tendencies toward voluntary restraint may reduce the cost of management, and certainly the cost of enforcement, since few fishers are likely to resist a regime that gives them more say in their own decisions. Other benefits also obtain: those to consumers, for example. When fishers work in a less-intense and unhurried manner, their fish are landed in better condition and over a longer season. And because capital costs are kept lower, the savings are passed along to consumers.

Social equity considerations are crucial when implementing this strategy, since to confer property rights is also to deny them to some who may have interests in the fishery. The problem is often best addressed by having recipients pay for the rights somehow, through rents, royalties, or whatever, to defray administrative and other costs. Though some critics oppose the institution of property rights on the grounds that they narrow participation and the distribution of benefits, this should not be seen as a fatal flaw. Property rights, after all, are the main means used to limit access to terrestrial resources around the world (FAO 1983: 19).

There are several common substrategies for instituting property rights in marine fisheries. Exclusive rights to harvest a certain species may be conferred, for instance, or fishers may obtain a license giving them other special rights. Fishers in particular communities may be granted territorial use rights in fisheries, or TURFS (see Panayotou 1984) or a share of a TAC.

Clearly, the common property, open-access system sets the stage for disaster in many fisheries. Until recently, many fisheries managers could not really address this core structural-institutional flaw because of political considerations, and thus were forced to find ever-more ingenious ways to mitigate its consequences. Unable to attack the problem at its source, they resorted to more politically acceptable strategies, such as closed areas and seasons, gear restrictions, and aggregate quotas. Yet even when the greatest talent, experience, and managerial effort, not to mention formidable expenditures, were brought to bear, the results were often disappointing. Depletion and overcapacity came about anyway.

Certainly throughout the first half of the twentieth century, there was an abiding hope among most fisheries managers that fisheries could be effectively managed as common property resources. For a long while, it seemed to be merely a matter of gaining enough experience to fine-tune the procedures. Only when, in the late 1960's, it became clear that in many fisheries the problems were growing, not abating, did it become less controversial to propose that the structural-institutional bases had to be changed. Practically everything else had already been tried.

By removing extensive sea territories from the realm of "the common heritage of mankind," the new EEZ system has opened the door for changing fisheries to other institutional bases. But, as an increasing number of nations are coming to see, without further internal changes, the new system has merely converted fisheries within their 200-mile limits to national rather than international commons.

Private-property regimes are much easier to implement in single-species fisheries than they are in developed fisheries which already have a multiplicity of fleet types as well as fish species. Ascribing property rights is also an effective means of deregulating a fishery, often effecting a greater economy of managerial effort by shifting much managerial responsibility downward, from central authorities to those holding the rights.

Among the various substrategies for conferring property rights, the institution of territorial use rights in fisheries (TURFS) is currently getting a lot of attention. Under this system, there tends to be a greater degree of local self-control or administration. However, to be truly effective over the long term, TURFS fisheries require regulation by higher governmental authority as well. Provision must be made for transfers of rights, and steps taken to prevent the development of monopolistic ownership.

Historically, ownership rights were ascribed to or asserted by individuals or small groups like the harbor gangs in Maine, discussed earlier. These days TURFS are usually granted to fishing communities, which in turn assign individuals exclusive rights to fish in certain areas or rights to take turns in particular areas. Indirect methods are also commonly employed to fine-tune the management regime, such as enforcing closed areas and seasons and imposing restrictions on fishing gear.

Critics of property-rights systems often point out that merely ascribing such rights to fishers does not automatically prevent overexploitation. In theory, they are no more inhibited from excess than the farmer who works his soil to exhaustion or the rancher who overgrazes his pastures. However, in practice this rarely seems to occur in these fisheries. Accumulating experience shows that fishers who enjoy property rights and can control access are not forced to compete so intensely for a "slice of the action," and they usually voluntarily maintain their fishing effort at levels that afford them reasonable profits, as well as reasonable sustained yields (Bell 1978: 137–38). This is true even in the face of high demand and prices, market forces that some critics insist will irresistibly tempt fishers into overexploitation. In many of the privately owned lobster and oyster fisheries around the world, for instance, fishers have maintained their effort at a level that guarantees good sustained yields, resisting the temptation to increase capacity in response to demand (Agnello and Donnelley 1975; Bell 1978: 147).

Apart from the tendency to promote voluntary restraint, the institution of a TURFS system often has other beneficial effects. Fishers are more likely to make expenditures for stock enhancement or to support community projects for the construction of improved landing sites, artificial reefs, and so forth (Beddington and Rettig 1984: 22).

The characteristics of the targeted species and the main means of capture are important considerations in establishing regimes emphasizing TURFS. These systems are especially appropriate for managing sedentary and semisedentary species like oysters and lobsters and for migratory species like salmon whose movements are predictable. They are also appropriate for managing species that are caught mainly through the use of fish-attracting and -aggregating devices, and are often particularly suitable where the capture technology is relatively immobile (e.g., fish traps and weirs).

Even when these requirements are met, some fisheries may not be suitable for TURFS management. Boundaries establishing rights must be capable of precise definition, enforcement must be feasible, and local communities must support the system if private rights are to work effectively (Christy 1982; Beddington and Rettig 1984: 23). Moreover, TURFS regimes are best instituted in developing fisheries. Switching to a private-property system would be extremely difficult and perhaps impossible in fisheries that are already mature. And the benefits are sure to be considerably weakened if the population is allowed to increase dramatically in the fishing area or production is tied too closely to modern markets. Thus, whatever the actual situation and the objectives underlying a proposed TURFS-management policy may be, implementing such regimes and assessing their effectiveness will usually be a very complicated matter.[5]

A quite different and also increasingly common means of instituting property rights in fisheries is the allocation of a TAC as a property right. Again, the expected benefit is that by guaranteeing fishers or fishing enterprises a proportion of the catch, the pressure for them to hurry or intensify their efforts will be removed. This evens out cycles in both employment and the fish market, thereby benefiting practically everybody—fishers and nonfishers alike. Again, there is less incentive to adopt more effective gear or larger, more powerful, and faster vessels. Safety is also promoted, since fishers will not feel as compelled to fish in bad weather.

If there is one conclusion to be drawn from the last two chapters, and indeed from those that preceded them, it is that no single management strategy is likely to work for any fishery, let alone all the fisheries in the world. Successful regulatory regimes must always be fitted to the particular type of fishery they seek to manage and must always employ a mixture of strategies. The appropriate mixture must be flexible enough to change as conditions change and must address not only the social, cultural, and economic concerns of fishers, but also the prevailing social equity and national policy concerns of their nations.

All management strategies, as we have seen, have certain costs and benefits, as well as certain problems and dilemmas; each and every one of these must be anticipated and evaluated before any course is decided on. Different countires will have different concerns, depending on their stage of development and their social, economic, and political institutions. Indeed, it may be that because of costs, the most appropriate course for some countries is to make no effort to manage the fisheries at all (FAO 1983: 1).

[5] Pollnac (1984) discusses complexities involved in investigating the effectiveness of TURFS management regimes.

Ultimately, the formulation of appropriate grand strategies for managing fisheries will always be a matter of public concern, an ongoing process of conflict and mediation aimed at providing the most good for the most people. Ultimately, too, we are left with the certain fact that fisheries management poses complex problems indeed; we will probably never find a panacea that will make it any easier. The best we can do is to fine-tune the existing regimes, consider other institutional bases for these renewable resources, experiment, and continue to learn from trial and error. We must also, finally, make the fishers themselves an important part of the process. For that is the only route that I know of to a truly humanized management of fisheries in the future.

ELEVEN

Future Management

In *Small Is Beautiful* (1973), a landmark work in the "appropriate technology" movement, E. F. Schumacher passionately urged that we re-develop our economic and resource-use systems so that human concerns are paramount. "We might do well to consider whether it is possible to have something better," he said, "a technology with a human face" (p. 146). It is a sentiment that many of us who are interested in the fisheries are inclined to share, particularly when we recall the glowing claims made for growth and modernization only a few decades ago. Now we look at many fisheries where such development took place and see local peoples who feel that they have been betrayed by the unforeseen, inadvertent, or uncontrollable consequences. In many instances, modernization and change backed local fishers into a corner, making them dependent on more transitory and complicated means of securing their livelihoods, without substantially improving their standard of living. This is why so many of the tenets that guided former modernization and development have now fallen into disrepute.

There is one timeless tenet, however, that has not fallen from grace, and it is an ancient one indeed. As the anthropologist Elman R. Service (1971: 34) stresses, the long-term survival of any organism or social system is jeopardized whenever it becomes too rigidly specialized. By his "Law of Evolutionary Potential," the more specialized the solutions a society develops for meeting its problems, the less flexibility it will have in responding to new problems arising in the future. Cultural extinction is thus a real possibility for societies that become overspecialized.

The lesson for fisheries management is abundantly clear: the more rigidly specialized and structured a management regime is, the less it will be able to cope with new eventualities. We must recognize that by trying to squeeze every conceivable benefit out of the fisheries, we are inevitably compelled to develop more complicated and specialized management regimes. Also, by striving for ever-increasing production, we run counter to the opinions of many contemporary economists and developmentalists, who warn that world economic growth cannot continue indefinitely, or at least not at the rate achieved since the beginning of the Industrial Revolution. Technological innovations, they argue, will probably not be able to sustain adequate levels of food production as they did in the past. As a result, in the exploitation of renewable natural resources, we are probably going to see a return to more labor-intensive approaches, a greater reliance on the productivity of natural systems, a decrease in productive effort, and the employment of less-intensive harvest technologies.

If we are to develop a more humanized process of fisheries management, we must promote three broad activities. First, we must incorporate indigenous means of management wherever practical and appropriate into fisheries management regimes. Second, we must incorporate more social analysis into the policy-making process. And third, we must find some way to give fishers themselves a greater say in their own domain.

Incorporating Indigenous Means

By incorporating local means of management into modern fisheries policies, we can address the need to deemphasize specialization and generalize the practice of fisheries management. That is not to say local ways of doing things are inevitably best, to be sure. We must first secure reliable data on their relative effectiveness and determine the extent to which exogenous factors may threaten their viability. Only then can we minimize the risk of making traditional practices important aspects of a modern management policy.

The really good thing about incorporating indigenous means of management, where feasible, is that this capitalizes on a natural process, rather than complicates, confounds, or opposes it. Considerable economies of managerial effort might thus be realized, as well as a reduction of conflict between fishers and fisheries managers.

Fisheries managers have traditionally regarded as particularly problematic the tendency of fishers to ignore or circumvent fisheries regulations.

Change in emphasis on compliance !

What is required here is a change of thinking, a more sympathetic under-standing that, instead of finding fault with fishers, finds fault with regimes that compel them to become lawbreakers.

Some caution is also in order here. While local means of management served many fishing peoples well for centuries, the time has long since passed in which such means can be relied on exclusively. Even prehistoric fishers sometimes found it impossible to behave as "enlightened preda-tors," and by now the rapid and extensive onslaught of modernization and industrialization, ever-increasing human population growth, ever-hungrier markets for food, and the greater mobility of fishers nearly everywhere make management solely by recourse to indigenous strategies out of the question in most fisheries.

Any still-unregulated fisheries today, and particularly those that are in-stituted as common property and permit open access, are ticking bombs— disasters waiting to happen. So, however objectionable it may seem, "top-down" management is here to stay. As Hardin (1968: 1247) said, "The social arrangements that produce responsibility are arrangements that create coercion, of some sort," by which he meant "mutual coercion, mu-tually agreed upon by the majority of people affected." Somehow, there-fore, we must find ways to make this sort of coercion more palatable in fisheries management.

Also problematic, as Berkes (1987: 90) cautions, is the fact that local management systems "do not mesh comfortably with government regula-tions," partly because "unlike scientific management, traditional manage-ment is not usually predicated on a biological rationale." Consequently, "reconciling traditional systems with scientific management will depend to a large extent on the ability of the bioeconomic model of fishery manage-ment, which has been changing rapidly since the 1970s, to accommodate different views of man-environment relations and to incorporate social concerns."

At the same time, we need to keep well in mind that the various means local peoples employ, or formerly employed, to regulate their fisheries have often been developed over long periods of time, after much trial and error, and that in many cases these peoples have a more comprehensive view of their fisheries and of how they should be managed than those fisheries managers whose perspectives have been unduly narrowed by immediate professional concerns and overspecialization. In this connection, Gary A. Klee, in the concluding chapter of his book *World Systems of Traditional Resource Management* (1980), emphasizes how well traditional cultures were adapted to their environments. "Their survival over thousands of years," he says, "is proof enough that they were good conservationists." In his view, "it would behoove all researchers . . . *to spend more time on recording and preserving traditional systems of resource management than devoting one's time to learning how to remove traditional obstacles to modernization*" (p. 283; emphasis in the original). Klee believes that collecting data on the effects of traditional conservation practices would yield new and important insights for fisheries management. He also believes this would not only help bring about more unified theories, tying together people and fish, but also lead to better predictions. "The available information on traditional conservation practices is meager and anecdotal," he laments, because past researchers neglected to look at indigenous means of resource conservation (p. 285).

I fully agree with Klee when he argues that it is time for fisheries managers to begin looking more closely at indigenous means of management and attempting to incorporate them into modern management regimes wherever possible. But I am skeptical of his insistence that the survival of traditional, preindustrial societies for thousands of years is proof that they were good conservationists. Emphatically, I feel that there was never anything about ancient societies as such that made them natural conservationists, their intimate associations with their marine ecosystems notwithstanding. Their long-term survival *may* be the result of their having been good conservationists. But it may merely owe to the fact that their populations were small, their harvest technologies rudimentary and relatively ineffective, and the aggregate demand for the marine resources they harvested fairly low. Whether or not they were in fact good conservationists is a matter that can be determined only by learning the particulars of their historical development. Recall, in this regard, the boom-and-bust cycles McEvoy (1986: 19–40) describes in his history of California's fisheries. Since this pattern developed long before the advent of colonialism and the integration of the local fisheries into capitalist economies, images of the aboriginal peoples' harmonious integration with their supporting ecosystems may be more illusory than real, perhaps as much a function of the point in the boom-and-bust cycle when they were first observed as anything else.

In any event, it would not be practical at this stage to leave very many local fishing societies alone to conduct their own fisheries management experiments as they wish. Few fishing peoples nowadays are so isolated as to be invulnerable to outside incursions, and fewer still remain unmoved by the temptations offered by modern markets. Therefore, it seems to me dangerous to give a blanket endorsement to indigenous means as such. Rather than joining those few fishers' advocates who insist that fishers always know best how to manage their fisheries, it seems to me more productive to take a conservative approach and simply isolate those indigenous management strategies that might be beneficially integrated into modern fisheries management regimes.

I hope that fisheries managers and policymakers will study the lessons of the two chapters on indigenous regulation as a guide to local strategies that may have applicability in modern management regimes. Once they have identified such strategies, they can then undertake quantitative assessments of the effectiveness of those strategies in the fisheries where they are still being used.

But those chapters are far from the final word on the subject. We need many more studies of indigenous management. The record from the Western Pacific Basin, more specifically Japan, Indonesia, Micronesia, Melanesia, and northern Australia, is particularly rich with examples of local self-management. As Ruddle and Akimichi (1984: 4) stress, in these regions "systems of sea tenure and their closely related conservation ethic probably attained historically a higher level of development than elsewhere." Moreover, they note, while Western-based colonialism and after that the imposition of Western modes of fisheries management eventually superseded indigenous systems of management in most of the region, Japan has retained a rich variety of local systems of sea tenure and "provides the world's best example of how traditional systems have continuously been adapted to changing circumstances and to fulfill modern functions," a conclusion with which I basically agree.

Yet surely fisheries in other parts of the world have much to teach us, too. And because local means of self-management were so often overlooked in earlier studies of fishing peoples, we should reexamine even formerly studied groups on the assumption that there is still much to learn from them.

Incorporating Social Analysis

We must rely to a greater degree on the expertise of anthropologists, sociologists, and other social scientists in the formulation of fisheries policies and management strategies. We must also challenge them to develop more standardized and quantitatively salient means of analyzing fisheries problems. In particular, more of these professionals should be employed by regulatory agencies. It is incredible, for example, that even now, more than a decade after the passage of legislation mandating that social analysis be an integral part of fisheries management, the National Marine Fisheries Service of the United States still has only one full-time sociologist (a cultural anthropologist) on its permanent headquarters staff.

As mentioned earlier, Fricke (1985: 47) suggests that policymakers and managers in the United States have been reluctant to rely on social and cultural studies in the formulation, implementation, and monitoring of fisheries policies largely for two reasons. The first is a general perception that, thanks to a lack of standardized methodologies and low quantitative salience, their recommendations are too risky to try out; the second has to do with the organizational climate in which fisheries policy and management are usually instituted.

The first problem places the burden for improvement squarely on the social scientists themselves. We simply must develop more rigorous techniques and the kind of data that will permit comparability, as well as integration, with other already formalized means of analysis.

What about the second problem? Here Fricke draws an interesting comparison between two agencies in the United States that are charged with regulating renewable resources: the Forest Service (FS) and the National Marine Fisheries Service (NMFS). The FS has a comparatively long history of responsibility for the country's renewable natural resources and has fully incorporated sociocultural and sociological analysis into the process of planning, implementing, and monitoring management policies for well over a decade. Nearly two dozen practitioners from the social sciences are employed as active members of interdisciplinary planning teams charged with preparing sociological-data baseline studies and with conducting social-impact assessments on an ongoing basis. Moreover, the FS regularly contracts the services of social scientists from universities and the private sector to assist in these efforts (Fricke 1985: 44).

By contrast, the NMFS relegates social analysis to a remote and fairly insignificant role. Organizationally, the NMFS splits the planning and policy-making responsibilities evenly between semiautonomous regional management councils and its central office. Though the regional councils do appoint sociologists and anthropologists to their scientific and statistical committees, their role, like that of the committees themselves, is basically advisory. Moreover, because of a tremendous work load, the regional councils seldom have time to explore adequately the social impact implications of policies being considered for implementation or revision. Thus, Fricke (p. 45) notes, in the central organization "review and advice on social impact analysis usually only occurs after the plan has been submitted for formal consideration by the agency." And even then, as we have seen, there is only one full-time sociologist available in the central office to participate in these reviews.

The organizational structure of the FS is quite different. It develops its programs internally, using interdisciplinary teams that include social analysts from both its central and regional offices. Moreover, those programs benefit from the expertise of social analysts at every key phase of the process, that is, in planning, implementation, and monitoring, so that the risks inherent in various management strategies are shared among the assorted members of the interdisciplinary teams (Fricke 1985: 40).

Though this organizational structure suggests a good model for incorporating social analysis into policy development for the fisheries, the structure would be far stronger if it also provided for the participation of the people who will be most affected in the policy-making process.[1] This may be easier said than done. It used to be fairly easy to identify fisheries users. They were, quite simply, the people who harvested marine resources for subsistence or commercial purposes. But in the modern era, increasingly diverse types of users have come to compete for those resources. Recreational users such as sport fishers must now be considered along with the more traditional fishers, and, in a larger sense, so must *all interests* laying claim to water resources—those who would use a fishery's waters for industrial plants, for agriculture, for urban water supplies, or for waste disposal, to name only a few. All such interested parties must somehow be granted a stake in establishing overall policies for managing water resources. From this perspective, a fishery's management policy must therefore be seen as merely part of a more comprehensive resources-management policy.

Michael L. Burton, G. Mark Schoepfle, and Marc L. Miller (1986: 261–62) make some excellent suggestions on how anthropologists in particular might go about studying problems of fisheries management and perhaps provide a more comprehensive view. Anthropology, they point out, integrates the theoretical perspectives of ecology, human cognition, and economics—all of which are important in the formulation of fisheries

---

[1] Wulff and Fiske, eds. (1987) contains essays suggesting the contributions anthropologists can make to the process of social-impact assessment. See also Korten and Klauss, eds. (1984) for many excellent suggestions on how planning for development can be made more people-centered.

management policies. The discipline has a tradition of considering biological and environmental data, a feature that is less common in other, closely related social science disciplines. Moreover, anthropology places a strong emphasis on ethnography, which among other things is concerned with identifying the relevant participants in a social system.

Particularly valuable is Burton, Schoepfle, and Miller's stress on studying natural resource management systems on four organizational levels: (1) the natural resources that are of interest to users, and the more thickly construed ecosystems in which they are found; (2) the associated social system or community of people who have interests in these resources, who possess cultural knowledge concerning their use, and who may have developed local political systems for their management; (3) local and regional managers and administrators of the resources, their advisers, and other such authorities; and (4) the ultimate decision makers in the overall political and economic system—national and international bureaucrats, politicians, corporate executives, and so forth—who decisively influence the establishment of overall policies dictating how the resources will be used and allocated. "Studying the whole system," they state, "requires that the conventional anthropological perspective on adaptation and social change be combined with a macroscopic view of global economic change" (p. 262).

Obviously this approach would be useful in the process of planning and implementing development programs in the fisheries, particularly when technological changes are being contemplated. As Pollnac (1982: 243) notes, "It is only by understanding the existing social organization and working within it that changes can effectively take place—changes not only in technology, but in aspects of social organization that inhibit change." Again, "social organization" must be more thickly construed to imply all the organizational levels at which human beings assert interests in certain resources.

Incorporating Fishing Peoples

Because the most pressing need in the fisheries today is to reconceptualize management policies in a way that makes human concerns paramount, it is essential that future management regimes provide for more extensive participation by the people concerned. What is most needed now is a shift away from autocratic and paternalistic modes of management to modes that rely on the joint efforts of traditional fisheries specialists and fishing peoples. This will not be an easy transition, but once it has become habitual and commonplace, many beneficial consequences should follow.

One way to begin that would have long-term benefits is to change educational systems in some manner that would require fishers to learn more about fisheries management, and fisheries managers more about fishers. It might be very beneficial, for instance, to require internships during the education of both groups, so that each would spend more time in the other's working environment. Cooperative co-management might develop thereafter as a natural extension of the process.

Regulatory agencies should also bring a greater number of expert fishers on staff, with authority to participate in decisions on management policies. Currently, the FAO is practically the only major agency concerned with the fisheries that does this. As Schumacher (1973: 158) observed, "The case for hope rests on the fact that ordinary people are often able to take a wider view, and a more 'humanistic' view, than is normally being taken by experts."

While many fishers' cooperatives and associations have been organized as a result of local initiatives—usually as a part of grass-roots movements—their raison d'être has most often been to secure more favorable marketing terms for their members (Jentoft 1985; McCay 1976). Other common organizational motives are to bring about better coordination between producers and marketers and to provide institutionalized foci for development programs. Only rarely have these organizations been formed for the explicit purpose of participating in fisheries management.

This is particularly lamentable now that there is a crisis in fisheries management in many parts of the world. The growing recognition that government-organized, or "top-down," management regimes are not working has led a number of theorists to propose various ways to diminish the state's role in management. Some would let market mechanisms regulate the fishing effort; others urge privatization. But as I indicated earlier, state involvement of some sort will continue to be necessary in most management regimes in the future, and privatization would be very difficult to institute where fisheries have traditionally been common property, open-access resources. Nevertheless, there is still plenty of latitude on just what the state's role will be, and delegating some of its authority to fishers by ensuring that they participate in the management process seems well worth further exploration (see Berkes 1986; Jentoft 1985: 328; and Kearney 1984).

Several new, often experimental approaches to this end are currently being tried out (see Bailey 1984b; Kearney 1984; and Pinkerton, ed.

[handwritten marginal note: role of fishers in fisheries management]

1989). Chief among them are cooperative co-management regimes, a strategy that has been encouraged and explored by international development agencies like the FAO for several years.[2] These regimes are not to be confused with the consultative management systems that have been in existence for several years in a number of countries, including Norway (Jentoft and Kristoffersen 1987, 1989), Canada (Kearney 1984), and the United States (Fricke 1985). Consultative management merely involves the establishment of government boards or committees that consult with fishers and their organizations before they establish fisheries policies and regulations, whereas in cooperative co-management fishers' organizations are not only consulted, but have a decision-making role. It is a mid-range strategy, falling evenly between management purely by market mechanisms at one extreme and management solely by direct state control at the other (Jentoft 1985: 331, n.d.: 12).[3]

In theory, when cooperative co-management is instituted in a fishery, its producers will perceive their mutual stake in sustaining resources at healthy and acceptable levels and therefore be motivated to police fishing effort among themselves. We would expect this in turn to reduce their conflicts with fisheries managers and also effect savings in costs associated with the managerial effort. Conflicts between fishers themselves should also be reduced if they all participate in making the important decisions on how the fishery's resources are allocated. As Courtland L. Smith (1988: 134) stresses: "For all users to feel the impact of their own actions on the whole, they must have some stake in the management of the resource. To develop incentives for resource conservation, harvesters must collectively experience feedback as to how their individual actions affect the resource." Smith adds that those who feel fishers cannot manage their own fisheries base their skepticism "on the current system of fishery management which promotes rather than reduces conflict," and which therefore also inadvertently elevates the costs of management (p. 136).

Several recent studies have shown just how effective cooperative co-management can be. Programs in both Iceland and Norway have been

---

[2] In 1982, I participated in a project whose goal was to explore how popular participation could be incorporated into fisheries development programs. The project was a part of the FAO's World Conference on Agrarian Reform and Rural Development.

[3] Pollnac (1987) reviews the main issues and problems involved in incorporating "popular participation" or "people's participation" in the fisheries policy process, particularly its important role in development projects. For other studies on cooperative co-management in the fisheries, see Berkes (1989a,b); Berkes and Kence (n.d.); Jentoft (1989); Kearney (1984); Korten, ed. (1986); McCay and Acheson, eds. (1987); Mollet, ed. (1986); Pinkerton (1986, 1988); and Pinkerton, ed. (1989).

pronounced successful (Durrenberger and Pálsson 1987b; Jentoft 1985; Jentoft and Kristoffersen 1987, 1989); and according to McCay (1980), there is a workable management arrangement in the New York Bight region of the mid-Atlantic coast, where fishers' cooperatives attend to many regulatory functions.

In a very different cultural milieu, Steve J. Langdon (1984a, b) describes how a Native American community in Alaska not only obtained legal sanction from the state to control access to locally important fishing territories, but also got the state to limit fishing in adjacent territories where local stocks might be harvested by outsiders. Similarly, in certain coastal communities in British Columbia, fishers and fisheries managers have developed a co-management system that permits commercial fishers to participate in the determination of desirable sustainable resource levels, allocations to users, and other aspects of management and decision making (Hilborn and Luedke 1987). In still other Pacific Northwest regions users' associations have shown considerable potential in helping to resolve long-standing disputes between small- and large-scale fishers, between fishers using different types of gear, and between fishers who prefer to fish during different seasons so that they can supply different markets (C. L. Smith 1988: 136–37).

The oldest and most successful co-management regimes are found in Japan, where cooperative organizations have long played an important role in fisheries regulation. Indeed, in many coastal fisheries, co-management is the rule rather than the exception (Jentoft n.d.: 9; Zengoryen 1984). But there, as in most of the Western nations with cooperative co-management regimes, the arrangement is confined entirely to coastal and inshore fisheries. In 1984, however, Britain began experimenting with co-management in some of its offshore fisheries, allowing producers' organizations to allocate TACs among themselves. Though this approach is still being tested, it has worked well enough so far for Goodlad (1986), chief executive of the Shetland Fish Producers' Organization, to describe it as "a successful experiment in the devolution of fisheries management responsibility from the National Government to the fishermen."

Still, cooperative co-management has not always been successful. British Columbia empowers certain local communities to manage their own fishing effort but pursues other policies that permit and even create a resource problem (Pinkerton 1988: 344). This points to the need for comprehensiveness wherever co-management policies are contemplated.

Despite promising results in the bulk of these experiments, most policy-makers remain cautious and skeptical about the approach. Appraising fisheries regulation in Canada and Norway, Jentoft (1985: 327) notes that instead of promoting more such regimes, these countries continue to rely on direct, exclusively state-instituted types of regulation. "The option of . . . cooperatively organized fishermen handling their own regulations has not been generally recognized as a management strategy." He finds this attitude particularly baffling in the case of Norway, whose most important cod fishery, the Lofoten fishery, has been successfully self-regulated ever since the Lofoten Act gave fishers the right of self-management in the 1890's. Important as this delegation of authority has been to the success of the fishery, it has not prompted much interest within the Norwegian government to see similar management regimes developed for the rest of the nation's fisheries.

Canadian officials are more forthrightly opposed to the idea. The Kirby Commission (1983), which was formed to study new alternatives for fisheries management, concluded that granting fishermen's cooperatives decisive power in fisheries management would lead to "economic and social chaos," and charged that the idea of co-management had not been developed in detail by those who advocate it. "[It] appears for the moment to be more of a catch-phrase than a well thought out proposal of substance" (pp. 9, 34). To a certain extent, the commission's fears and reservations are well founded in the light of the experience in the Bay of Fundy herring fishery; after the Canadian government delegated management responsibilities to fishing cooperatives there, resource levels declined and the fishers' organizations became rife with internal disputes (Kearney 1984).

Cooperative co-management approaches thus have the potential of generating new problems or heightening existing ones. Internal conflicts and disputes are only one possible outcome. Users' organizations may also run into trouble if they are given insufficient autonomy and decision-making powers. And unless they receive sufficient technical assistance, they are often not competent to undertake the management responsibilities that have been delegated to them (Jentoft n.d.: 17).

Unfortunately, cooperative co-management approaches are sometimes opposed for less legitimate reasons. Powerful economic interests may see them as threats to their continued economic dominance, and management personnel may fear that they will infringe on important professional turf. After all, cooperative co-management brings many newcomers into a management scheme, and since there are only finite resources for supporting the management effort, some entrenched professionals may eventually be forced out.

McCay's (1988) excellent case study, presented "from an anthropological perspective," of an experimental effort to bring about co-management in New Jersey's clam fisheries contains important lessons for anyone who may become involved in planning and implementing a co-management effort.

What prompted the experiment was the increasing scarcity of the hard clams targeted by the bay fishers. Given an "almost ritualistic hostility . . . between state officials and baymen" that was "generated by a management style in which the state developed its plans without involvement of those affected by them" (p. 329), McCay and her colleagues designed an experiment in which the two groups, joined by scientists concerned with the fishery, would work together to create spawner sanctuaries. In the event, the experiment had only limited success in revitalizing the clam stocks, but it proved empirically valuable in revealing complex "socio-cultural interactions among a variety of constituencies and individuals who attempted to 'muddle through' the problem solving process together" (p. 327).

The experiment was no sooner launched than various problems arose that confounded the effort. McCay attributes these problems mainly to "our underestimation of the effects of long-term factionalism within the shellfisheries of the region and our overestimation of the willingness or ability of state 'co-managers' to cooperate" (p. 334). Groups of fishers who had long been competitors were reluctant to work together cooperatively, and in hindsight it became clear that they should have been better integrated into the project before it was started. Moreover, according to McCay, state officials were constrained in their ability to cooperate because of "the tendency of individuals working for the state to minimize any action that will cause a reaction." As one state employee put it, "I didn't want to aggravate them [i.e., local fishers] because they would just turn around and aggravate me" (p. 335). In these circumstances, only the rare scientist or administrator was inclined to make a strong commitment to the grass-roots approach to the fishery's management problems:

The people who made this project work are those who were able to cast scientific knowledge not only into ordinary language but also into the practical concerns of ordinary people; they are people who, whether scientist or bureaucrat or clammer, seemed to care little about competing claims for legitimacy but instead to be most concerned about what the social scientists call "praxis" or "social action," and what others might describe as "getting something done" (McCay 1988: 337).

In a fishery like this, where the fishers who compete for the same resources are dispersed and not organized, managers and scientists must take the lead in addressing the stubborn ecological and management problems that have not responded to more traditional management approaches. This is why McCay insists that "scientists increasingly must be able to interact with others, of different disciplines, and of different professional and social backgrounds. . . . They must be able to appreciate the points of views of others, which include different criteria of quality and truth" (p. 337).

Scientists and professional managers should not fear co-management approaches merely because they are awkward to implement, or because their future consequences are difficult to assess. And those who try them out should not become too discouraged when their first attempts are less than completely successful. Indeed, it could be that "muddling through" is the most appropriate strategy for tackling difficult management problems that have not responded to attempts to solve them by other means (see Lindblom 1959, 1979).

Humanizing Fisheries Management

The problems of fishers, particularly small-scale fishers, are so complex that they have taken up most of this book. Hence I have not said much about closely related groups like processors and marketers.[4] Nor have I said enough about the women in fishing communities, who are often the bulwarks of the local social order, as well as the main marketers of the catch. Obviously, all the people who are ancillary to fish production must be integral considerations in a fishery's management policy, since they are directly linked, both socially and economically, to the primary producers, and only the need to bound this discussion somewhere has caused me to rush over these subjects. For the same reason, I have not discussed aquaculture, only some types of which are in any case pertinent to marine-capture fisheries. Mariculture projects, which basically involve the improvement of marine ecosystems in order to increase their productivity,

[4] Peterson and Georgianna's 1988 article provides rich insights into the complex problems inherent in a modern fish market. Among their citations are four works of particular interest to students of fish-marketing systems: Georgianna and Hogan (1986); Peterson (1973); D. J. White (1954); and Wilson (1980).

plainly have an important bearing on fisheries and fisheries policies. But man-made aquatic environments such as fish ponds are more analogous to farming than to fishing as such. This type of aquaculture usually entails private resources and poses several problems very different from those posed by marine-capture fisheries. A full consideration of the subject would thus inevitably take us very far afield indeed.[5]

That said, let me return to the book's main concern, small-scale fishers, and specifically now those in the developing nations, many of whom are experiencing severe problems because their fisheries have been transformed to producing for export. In the rural shrimp fishing community of Pacific Mexico that I have studied for many years, for instance, late summer often finds many of the local inhabitants close to starvation. There are no agricultural harvests, and the government closes the most important inshore fishing grounds to protect juvenile shrimp for later harvest by exporters. Furthermore, nearly all the local fishers are excluded from participating in the export industry because their "artisanal" mode of production is seen as too unproductive, inefficient, and extensive.

Consequently, as the summer wears on and local food supplies dwindle, adult community members become irritable and restless, the incidence of violent crime increases, and some children eat dirt or sand to allay the gnawing hunger in their bellies. There also seems to be a higher incidence of death among old people and infants during this season.

Meanwhile, the surrounding estuaries and lagoons teem with juvenile shrimp. Local fishers cannot forgo poaching during the closed season in order to feed themselves and their families, even though they risk capture, fines, and imprisonment at the hands of the soldiers the government deploys to enforce the prerogatives of the shrimp-export industry.

Newspapers in nearby urban centers are wont to refer to these people as "internal pirates," "contrabanders," and "thieves of the national patrimony," and fisheries managers to decry the additional pressures they place on an already highly pressured resource. But the problem in this fishery does not stem from any flaw in the local fishers' characters or from a pure disregard for the law. Instead, it stems from the development of a management policy that long ago favored the export industry at their expense.

From a national economic perspective, managing the fishery this way makes good sense: the economic benefits are maximized (nearly approach-

---

[5] An excellent and comprehensive source book on aquaculture is Bardach et al. (1972), and for an interesting collection of essays on the problems of promoting it in developing nations, see L. J. Smith and S. B. Peterson, eds. (1982).

ing MEY), and considerable foreign exchange is earned. But from a local perspective, the policy is a disaster and absolutely inhumane (McGoodwin 1987). In essence, impoverished maritime peoples are prevented from harvesting an important local food resource so that it may become a discretionary, luxury food item for other people far away, who already have surplus food supplies. This is the world turned upside down, and it is difficult to imagine how anyone who has seen at first hand the grievous problems the policy causes could make a case for continuing it.

In fisheries like this one, meeting local subsistence needs should be considered the first tier of management policy—a prime part of the fishery's fixed overhead. Only after the local people's food supply is ensured should wider means of distribution be explored. For as Bell (1978: 148) observes: "If a society is on the verge of starvation, its members could not care less about overfishing. There is no tomorrow for such a society, so they may be persuaded to take sizable chunks of the entire fishery biomass for food. We say they have a short time preference: they would rather consume today than save the fish for harvest sometime in the future."

These days we are so bombarded by newspaper stories and television documentaries about famine and poverty that we must guard against becoming desensitized. We must also guard against automatically assuming that just because there are food shortages, there has been a failure of local food production. As Frances Moore Lappé and Joseph Collins (1977) remind us, we must often look beyond the myth of scarcity. Quite often food crises are due more to the inequitable distribution of supplies than to a shortage.

Similarly, subsistence economies should not be categorically seen as backward and underdeveloped. If they have sustained their peoples for a long time, that is more than can be said of many modern economies, periodically beset as they have been by violent economic swings, if not extreme reversals and even collapse during worldwide depressions. Indeed, it is the subsistence economies that have shown themselves to be the most immune to the effects of such disasters.

Fisheries that merely feed local peoples may appear backward and underdeveloped, but those that concentrate on feeding the affluent in distant lands while denying food to their own impoverished population merit far more opprobrious adjectives. When fishers provide their own food in a way that gives them satisfaction and affirms the core aspects of their traditions and culture, they sustain themselves in a manner that has timeless value for them and simultaneously makes a positive contribution to society at large. On the other hand, when they are stripped of their ability to provide for themselves, losing at the same time their unique cultural identities, they become unhealthy, unproductive, and dependent. Plagued by anomie and disaffection, they add to the suffering and waste of human potential on this planet.

Old philosophies die hard. Allowing a fishery to be developed until it reaches its bioeconomic equilibrium is still advocated by a few social philosophers and economic theorists—mainly those who insist that the government that governs least governs best. On their argument, allowing the brute forces of economics full sway will more quickly resolve fisheries problems than will interventionist measures. In essence, they feel that the kind of management we see today only perpetuates the fisheries' problems. This view may have some validity for addressing other sorts of economic problems, but it is utterly naïve as a solution for the fisheries. Considerable accumulated experience has already shown that permitting an open-access, common property fishery to reach its bioeconomic equilibrium practically guarantees the perpetuation of resource depletion and low productivity, economic marginality, stagnation, inefficiency, and human misery.

If we are to manage our fisheries in the wisest possible way, we need to devise formal frameworks that comprehensively construe the social, economic, and biological consequences of various management policies. Here, bioeconomic models may continue to lead the way, but more sophisticated means of social analysis must be developed and incorporated into them. In this regard, social scientists must not only develop more rigorous means of analysis; they must also better inform themselves about fisheries science, its historical development, and its current problems and practices. As R. Bruce Rettig (1987: 405) notes, "Social scientists must recognize that they are not creating information in a vacuum; rather they are adding to, verifying, and correcting a vast amount of common knowledge about fisheries."

Thus far, the science of fisheries management has reached its highest level of sophistication in dealing with large-scale fishers and fisheries. But even there the frameworks that are commonly employed suggest a still-inexact science. The levels of analysis commonly applied in small-scale fisheries are not nearly so advanced. As Kesteven (1976: 134) stated over a decade ago, "I know of no systematic and comprehensive study of an artisanal fishery in which catch and effort have been measured, mortalities estimated, and the value obtained for the exploitation index." Since then, more sophisticated analytical frameworks have been proposed, and some progress has been made, but considerably more work of this type still needs to be undertaken.

Future Development

In the postwar era, scholars and developmentalists enthusiastically sought to promote fundamental social and economic changes in the developing nations, nearly all of which were inhabited by subsistence-oriented peoples. The developed, industrialized nations were held up as exemplary models, and there was often a rather pervasive and convenient amnesia about just how rocky the actual road to development had been in most of them.[6] Implicit in these efforts was the assumption that the target nations' social, cultural, and economic systems were inherently backward and had to be changed. Here was imperialism in a new guise—cultural imperialism, the idea that the underdeveloped nations had to become more like the developed ones. Maintenance of the status quo in the underdeveloped nations was unthinkable because the status quo itself was seen as a problem to be resolved.

Moreover, even when the aid-giving nations' motives were of the purest—which was by no means always the case—their "guided" development programs often left new human suffering in their wake. Just as the Green Revolution succeeded in increasing agricultural production in most of the targeted countries but often made matters worse for a large chunk of their populations, so what Bailey (1985) terms the Blue Revolution has spelled disaster for many of the developing countries' fishers. Especially by encouraging the development of export markets for highly valued seafoods—shrimp and tuna, among others—international development organizations often inadvertently lent support to national fisheries policies that favored capital-intensive industrial fishing at the expense of the small-scale fisher. Quite often, thanks to the assistance of foreign donors, powerful interests in a country were able to acquire new technologies and utterly squeeze out small operators.

Elsewhere, countries embarked on their own development programs, similarly promoting the breakdown of traditional patterns of exploitation

---

[6] A particularly comprehensive and historically detailed account of the rocky road to "development" in a developed nation's fisheries is McEvoy's tour de force, *The Fisherman's Problem: Ecology and Law in the California Fisheries, 1850–1980* (1986).

in countless small-scale fisheries, most of which had been relatively extensive and low yield. Traditional values emphasizing sharing and reciprocity thus gave way to ruthless competition, and various types of traditional fishing vessels and gear that had kept large numbers of people employed were rendered obsolete or uncompetitive (see, e.g., Bailey 1984a; Cordell 1974, 1978; Fraser 1966; McGoodwin 1987; Neal 1982; Thomson 1980).

In view of these unfortunate experiences, it is essential that the planners of future development projects not only better anticipate the impact and consequences of their contemplated action but also strive for more sustainable forms of development (see Clark and Munn, eds., 1986). Development projects must be coherently planned, so that advances in one sector will not undermine the well-being of another. Care must be taken in encouraging timber-cutting industries, for example, lest they lead to widespread erosion and harmful siltation in fisheries farther downstream. Likewise, new mining operations and industrial plants, the expansion of agriculture, and urbanization can place great demands on, and pollute, water resources.

While fisheries development policies will continue to be determined by the developing nations themselves, bilateral and multilateral assistance agencies can still importantly influence the process. As Conner Bailey, Dean Cycon, and Michael Morris (1986: 1273–74) note, these agencies are in a good position "to engage in policy dialogue with national decision makers to shape fisheries management and development strategies that promote both sustainable harvests and social justice." In a related vein, Franke and Chasin (1980: 130), speaking of the severe problems experienced in the African Sahel, argue: "If the economic and political relations have within them the causes of the problem, then it is those relations that must be altered. The altering of these relationships is primarily a task for the Sahelian peoples themselves, while the role of outside 'experts' becomes one of encouraging support for the Sahelians' efforts among the people of our own society." Indeed, development "experts" are in a good position to promote a more ecumenical process of policy formulation by urging policymakers to draw local representatives of the people who are most likely to be impacted into the process, and by offering not only material and technical assistance, as they have in the past, but also the assistance of experts in such domains as socioeconomic-impact analysis.

Fisheries development projects must never be considered apart from their implications for fisheries management, since changes in exploitation levels and allocation patterns nearly always necessitate changes in a management regime. As things stand, in many developing nations the fisheries development and fisheries management agencies are parts of different bureaucracies. Thus the management agencies have sometimes found themselves forced to cope with new problems growing out of actions taken by their development-minded counterparts. Obviously, if indiscriminate actions taken now to encourage the growth of a fishery are not to confound the possibilities for successful management later on, related agencies within a national government, as well as any foreign agencies assisting them, must work together in a more integrated and coordinated fashion (FAO 1983: 24).

It is not going to be easy to bring about the necessary structural changes in the social, political, and economic systems of developing nations—much less in the world's overall political economy. As we have come to realize, too late in most cases, many of the original, predevelopment social and economic systems had more to offer than was originally thought. Unfortunately, now that they have been uprooted, there seems to be no turning back. There is one positive aspect to all this, however, and it is that similar developmental disasters in the future may be avoided by remembering the negative lessons of the recent past. Already many developing nations are exploring other means for improving their people's social and economic welfare and becoming very selective about what types of assistance they will take from the developed nations. A few are even refusing such help altogether.

Fisheries Policies for the Future

To this point, Maiolo and Orbach (1982: v) note, the study of fisheries management and policy has been "either the study of conservation in the biological sense or the study of politics." What is so badly needed now, they stress, is to reconceptualize fisheries management and policy as "a complex process of human behavior and interaction characterized as much by the social and cultural values of the participants as by the compelling drive for conservation of marine biological species or the achievement of political purpose."

Grand strategies for alleviating the most common problems in fisheries today must aim at limiting the overall demand for seafoods and reversing the degradation of marine environments, a trend that is fueled not only by this demand, but also by other, competing demands on water resources. There is also a great need either to refine common property, open-access regimes or to do away with them. Tendencies toward overcapitalization must also be checked somehow, particularly in the many fisheries where capitalism has seemingly become sated with itself.

Future policies must take much better account of fishers as well. Where fishers do not have the same rights as any other group of workers, that situation should be redressed. Future policies should also promote the development of organizations that give fishers greater influence in fish markets and allow them to participate more fully in the establishment and monitoring of fisheries management regimes. Means must also be found to help reduce fishers' estrangement from their families and communities. So, however tempting it may seem from other perspectives, development programs should weigh the social costs of enabling fishers to stay at sea longer, enticing them away from day trips, for example. More broadly, community development programs should be instituted that provide vitally needed services, such as psychological counseling and alcoholism programs for fishers and social services for their families. More stringent provisions for the personal safety of fishers while they are at sea should also be an essential aspect of future fisheries policies, since fishing is still one of the most dangerous occupations in the world.

Though fishing peoples tend to be colorful and of exotic interest to the general public, they also tend to be held in low esteem. A concerted effort should be made to combat this attitude, say through public information programs stressing fishers' contributions to human food supplies. Attempts should also be made to counter the negative view of fishers prompted by certain environmental groups, including giving fishers "equal time" to show their sensitivity to environmental problems.

The public should also be made more aware of the costs of water pollution and of the need to make polluters pay for damages to the marine fisheries that serve us all. Also, where peoples in coastal communities support the development of tourism and recreational fishing, they must be allowed to guide and participate in that development to the fullest extent possible.

Of course, nobody can predict the future, not even the immediate future. At the most pessimistic extreme, doomsdayers stress the increasing likelihood of nuclear war or see the planet eventually rendered unfit for human life by pollution. Others predict an increasing incidence of famines or of diseases like AIDS against which most of humanity will have no defense. Still others foresee the spread of terrorism and anarchy, prompting more and more oppressive and autocratic governmental regimes. Even many fairly moderate soothsayers are inclined to look pessimistically on the soaring world population and the widening gap between food needs and food supplies. Hardin (1988: 5), for instance, warns that "population policy for human beings must be governed by a commitment never to transcend the cultural carrying capacity."

Currently, the fisheries provide around 13 percent of the total animal protein available for human consumption, although in many parts of the world, particularly in East and Southeast Asia, fish supplies nearly all of the animal protein consumed, particularly among the poor segments of the population (E. N. Anderson, Jr., 1975). Expecting the fisheries to make a significantly greater contribution in the near future seems a vain hope. The world fish catch, after leveling off a couple of decades ago, has been growing at a painfully slow rate ever since. Thus, while many fisheries analysts hope that the catch will grow somewhat faster with better management and more effort directed toward harvesting underutilized species, they also warn that the soaring human population and the soaring demand for fish products, the proliferation of increasingly more effective fishing technologies, and the ongoing degradation of marine environments, make it uncertain that even today's production levels can be sustained in the future. Indeed, it seems equally likely that the world's fish catch may soon begin to decline (Idyll 1978).

Frederick W. Bell's book *Food from the Sea* (1978), an important critique of contemporary fisheries management, ends on this gloomy note:

In conclusion, we are very pessimistic about the future of food from the sea. Governments are openly pursuing policies to diminish the resource base through overcapitalization. International commissions are ineffective. . . . The fishery resources of the world are becoming more and more valuable . . . and will be enjoyed, as in the past, by the more affluent societies. In addition, many of the underexploited species (krill, lanternfish) may be used more often than not for reduction purposes, to serve as feed for higher-value protein. As for the less-developed countries, extended fishery jurisdiction does offer them a chance to collect rents for the use of their fishery resources and thus to earn foreign exchange. It offers less chance to improve their direct food consumption (pp. 357–58).

Achieving stability in the fisheries while also maximizing social and economic well-being promises to be a formidable task in this era of rapid change. It will be difficult indeed to operationalize OSY in a world that turns its back on its past. Under the circumstances, it is particularly important that we show a redoubled concern for small-scale fishers, many of whom still practice old ways that could teach us much about how to proceed.

For my own part, I feel there is good reason to maintain at least a cautiously optimistic point of view. Marine organisms are renewable resources, after all, and even when they are exploited to the point of depletion, they can usually rebound if permitted to do so. Moreover, many of the world's inhabitants are now considerably more disenchanted with industrialization and modernization than they were only a few decades ago, so there may be less support for development for the sake of development in the future. Other trends and movements, some of which are just now gathering steam, are also cause for hope, including the trend toward economic decentralization, the "appropriate technology" movement, and especially the worldwide environmental movement, which is garnering increasing political support.

We may perhaps take some comfort in the fact, finally, that it has been less than a hundred years since the crises in the North Atlantic and North Seas fisheries first caused alarm, and though it took several decades to bring those crises under control, they *were* brought under control. We must be patient in our hope to find a solution for the future. Meanwhile, even if the world's fisheries are not all that we would wish, there are still plenty of fish in the sea, plenty of fishers eager to harvest them—and a growing number of students of the fisheries concerned to see the fisheries maintained in good health.

Reference Matter

References

Abrahams, Roger D. 1974. *Deep the water, shallow the shore.* Austin: The University of Texas Press.

Acheson, James M. 1972. The territories of the lobstermen. *Natural History* 81(4): 60–69.

———. 1975. The lobster fiefs: economic and ecological effects of territoriality in the Maine lobster industry. *Human Ecology* 3: 183–207.

———. 1977. Technical skills and fishing success in the Maine lobster industry. Pp. 111–38 in H. Lechtman and R. Merrill, eds., *Styles, organization, and dynamics of technology.* St. Paul, Minn.: West.

———. 1979. Variations in traditional inshore fishing rights in Maine lobstering communities. Pp. 253–76 in Andersen, ed.

———. 1981a. Factors influencing production of metal and wooden lobster traps. Technical Report no. 63. Orono: University of Maine Sea Grant Publication.

———. 1981b. Anthropology of fishing. *Annual Review of Anthropology* 10: 275–316.

———. 1982. Metal traps: a key innovation in the Maine lobster industry. Pp. 279–312 in Maiolo and Orbach, eds.

———. 1987. The lobster fiefs revisited: economic and ecological effects of territoriality in Maine lobster fishing. Pp. 37–65 in McCay and Acheson, eds.

———. 1988a. *The lobster gangs of Maine.* Hanover, N.H.: University Press of New England.

———. 1988b. Patterns of gear changes in the Maine fishing industry. *Mast* 1: 49–65.

Agnello, Richard J., and Lawrence P. Donnelley. 1975. Property rights and efficiency in the oyster industry. *Journal of Law and Economics* 13: 521–33.

Akimichi, Tomoya. 1984. Territorial regulation in the small-scale fisheries of Itoman, Okinawa. Pp. 89–120 in Ruddle and Akimichi, eds.

Akimichi, Tomoya, and Kenneth Ruddle. 1984. The historical development of territorial rights and fishery regulations in Okinawan inshore waters. Pp. 37–88 in Ruddle and Akimichi, eds.

Alexander, Paul. 1975. Do fisheries experts aid fisheries development? The case of Sri Lanka. *Maritime Studies and Management* 3(1): 5–11.

———. 1977. Sea tenure in southern Sri Lanka. *Ethnology* 16: 231–51.

Allison, Charlene J., Sue-Ellen Jacobs, and Mary A. Porter. 1990. *Winds of change: women in northwest commercial fishing*. Seattle: University of Washington Press.

Alverson, D. L., and G. J. Paulik. 1973. Objectives and problems of managing aquatic living resources. *Journal of the Fisheries Research Board of Canada* 30: 1936–47.

Andersen, Raoul. 1972. Hunt and deceive: information management in Newfoundland deep-sea trawler fishing. Pp. 120–40 in Andersen and Wadel, eds.

———. 1974. North Atlantic fishing adaptations: origins and directions. Pp. 15–33 in Pontecorvo, ed.

———. 1975. Optimal sustainable yield in inland recreational fisheries management. Pp. 29–38 in Roedel, ed.

———. 1976. The small island society and coastal resource management: the Bermudian experience. Pp. 255–77 in Douglas M. Johnston, ed., *Marine policy and the coastal community: the impact of the Law of the Sea*. London: Croom-Helm.

———. 1979. Public and private access management in Newfoundland fishing. Pp. 299–336 in Andersen, ed.

———. 1980. Hunt and conceal: information management in Newfoundland deep-sea trawler fishing. Pp. 205–28 in S. K. Tefft, ed., *Secrecy*. New York: Human Sciences Press.

———. 1982. Extended jurisdiction and fisherman access to resources: new directions, new imperatives. Pp. 17–34 in Maiolo and Orbach, eds.

———. 1988. Usufruct and contradiction: territorial custom and abuse in Newfoundland's banks schooner and dory fishery. *Maritime Anthropological Studies* 1: 81–102.

———, ed. 1979. *North Atlantic maritime cultures: anthropological essays on changing adaptations*. The Hague: Mouton.

Andersen, Raoul, and R. Geoffrey Stiles. 1973. Resource management and spatial competition in Newfoundland fishing: an exploratory essay. Pp. 44–66 in Fricke, ed.

Andersen, Raoul, and Cato Wadel, eds. 1972. *North Atlantic fishermen: anthropological essays on modern fishing*. St. John's: Institute of Social and Economic Research, Memorial University of Newfoundland.

Anderson, E. N., Jr. 1975. Chinese fishermen in Hong Kong and in Malaysia. Pp. 231–46 in Casteel and Quimby, eds.

Anderson, Lee G. 1977. *The economics of fisheries management*. Baltimore, Md.: Johns Hopkins University Press.

Anonymous. 1924. The Greenwich pensioner. Pp. 209–10 in John Masefield, ed., *A sailor's garland*. New York: Macmillan.

Anonymous. 1988. Warning: how to lose a family. *National Fisherman*, Sept.: 7, 58.

Anson, P. F. 1965. *Fisher folk-lore*. London: Faith Press.

Apostle, Richard, Leonard Kasdan, and Arthur Hanson. 1985. Work satisfaction and community attachment among fishermen in southwest Nova Scotia. *Canadian Journal of Fisheries and Aquatic Sciences* 42: 256–67.

Aronoff, J. 1967. *Psychological needs and cultural systems*. Princeton, N.J.: D. Van Nostrand.

Bailey, Conner. 1984a. Fisheries resource conflict and political resolution: Indonesia's 1980 trawl ban. Paper presented to the annual meeting of the Rural Sociological Society, Aug. 22–25, Texas A&M University.

———. 1984b. Managing an open access resource: the case of coastal fisheries. Pp. 97–103 in Korten and Klauss, eds.

———. 1985. Blue revolution: the impact of technological innovation on Third World fisheries. *The Rural Sociologist* 5(4): 259–66.

———. 1986. Government protection of traditional resource use rights—the case of Indonesian fisheries. Pp. 292–308 in Korten, ed.

———. 1987. Social consequences of excess fishing effort. *Proceedings, Symposium on the Exploitation and Management of Marine Fishery Resources in Southeast Asia, Darwin, Australia, February 16–19*. Bangkok: Regional Office for Asia and the Pacific, Food and Agriculture Organization, RAPA Report 1987/10.

———. 1988a. The social consequences of tropical shrimp mariculture development. *Ocean and Shoreline Management* 11(1): 31–44.

———. 1988b. Optimal development of Third World fisheries. Pp. 105–28 in Michael A. Morris, ed., *North-South perspectives on marine policy*. Westview Special Studies in Ocean Science and Policy. Boulder, Colo.: Westview Press.

Bailey, Conner, Dean Cycon, and Michael Morris. 1986. Fisheries development in the Third World: the role of international agencies. *World Development* 14: 1269–75.

Baks, Chris, and Els Postel-Coster. 1977. Fishing communities on the Scottish east coast: traditions in a modern setting. Pp. 23–40 in M. E. Smith, ed.

Baldwin, J., and Stewart Brand. 1978. *Soft tech*. New York: Penguin.

Ballonoff, Paul A. 1974. *Mathematical models of social and cognitive structures: contributions to the mathematical development of anthropology*. Urbana: University of Illinois Press.

Baranov, F. 1918. On the question of the biological basis of fisheries. *Nauchnyi Issledovatelskii Ikhtiologecheschii Institut Isvestia* 1: 81–128.

Bardach, John E., John H. Ryther, and William O. McLarney. 1972. *Aquaculture: the farming and husbandry of freshwater and marine organisms*. New York: Wiley Interscience.

Barth, Fredrik. 1966. *Models of social organization*. Occasional Paper no. 23. London: Royal Anthropological Institute of Great Britain and Ireland.

Bartlett, John. 1968. *Familiar quotations*, 14th ed. Boston: Little, Brown.

Beals, Ralph, and J. Hester. 1974. *California Indians*, vol. 1. New York: Garland.

Beddington, John R., and R. Bruce Rettig. 1984. *Approaches to the regulation of fishing effort.* FAO Fisheries Technical Paper no. 243. Rome: FAO.

Befu, Harumi. 1981. Political ecology of fishing in Japan: techno-environmental impact of industrialization in the inland sea. *Research in Economic Anthropology* 3: 323–37.

Bell, Frederick W. 1978. *Food from the sea: the economics and politics of ocean fisheries.* Boulder, Colo.: Westview Press.

Ben-Yami, Menachem. 1980. Community fisheries centres and the transfer of technology to small-scale fisheries. Pp. 936–48 in *Proceedings of the 19th session of the Indo-Pacific Fisheries Council, Kyoto, Japan, May 21–30.*

Berkes, Fikret. 1977. Fishery resources use in a subarctic Indian community. *Human Ecology* 5: 289–307.

———. 1985. Fishermen and "the tragedy of the commons." *Environmental Conservation* 12(3): 199–205.

———. 1986. Local level management and the commons problem. *Marine Policy* 10: 215–29.

———. 1987. Common-property resource management and Cree Indian fisheries in subarctic Canada. Pp. 66–91 in McCay and Acheson, eds.

———. 1989a. Local-level resource management studies and programs: the Great Lakes region and Ontario. Pp. 95–118 in F. G. Cohen and A. J. Hanson, eds., *Community-based resource management in Canada: an inventory of research and projects.* Report no. 21. Ottawa, Canada: Man in the Biosphere Programme, United Nations Educational, Scientific, and Cultural Organization.

———. 1989b. Co-management and the James Bay Agreement. Pp. 189–208 in Pinkerton, ed.

———, ed. 1989. *Common property resources: ecology and community-based sustainable development.* New York: Columbia University Press.

Berkes, F., D. Feeny, B. J. McCay, and J. M. Acheson. 1989. The benefits of the commons. *Nature* 340: 91–93.

Berkes, Fikret, and Aykut Kence. n.d. Fisheries and the prisoner's dilemma game: conditions for the evolution of cooperation among users of common property resources. *Middle East Technical University Journal of Pure and Applied Sciences* 20 (2). In press.

Berleant-Schiller, Riva. 1982. Development proposals and small-scale fishing in the Caribbean. Pp. 115–39 in Maiolo and Orbach, eds.

Bernard, H. Russell. 1967. Kalymnian sponge diving. *Human Biology* 39(2): 103–30.

———. 1972. Kalymnos, island of the sponge fishermen. Pp. 277–316 in H. R. Bernard and P. Pelto, eds., *Technology and social change.* New York: Macmillan.

———. 1976. Is there an anthropology for everyone? *Reviews in Anthropology* 3(5): 478–85.

Besançon, Jacques. 1965. *Géographie de la pêche.* Paris: Gallimard.

Beverton, R. 1953. Some observations on the principles of fishery regulation. *Journal du Conseil International pour l'Exploration de la Mer* 19: 56–68.

Beverton, R., and S. Holt. 1957. On the dynamics of exploited fish populations. *Fisheries Investigations Series* 2(19). London: Fisheries and Food Department of the Ministry of Agriculture.

Bishop, C. 1981. Northeastern Indian concepts of conservation and the fur trade: a critique of Calvin Martin's thesis. Pp. 39–58 in Krech, ed.

Blair, Emma Helen, and James A. Robertson, eds. and trans. 1905. *The Philippine islands.* 55 vols. Cleveland, Ohio: A. H. Clark.

Blair, H. W. 1974. *The elusiveness of equity: institutional approaches to rural development in Bangladesh.* Ithaca, N.Y.: Rural Development Committee, Center for International Studies, Cornell University.

Bowles, Francis P. 1973. Natural regulation of an island fishing community. Ph.D. diss., Harvard University.

Brewer, Gerald D. 1986. Methods for synthesis: policy exercises. Pp. 455–73 in W. C. Clark and R. E. Munn, eds.

Brightman, Robert A. 1987. Conservation and resource depletion: the case of the boreal forest Algonquians. Pp. 121–41 in McCay and Acheson, eds.

Britan, G. 1979. "Modernization" on the North Atlantic Coast: the transformation of a traditional Newfoundland fishing village. Pp. 65–81 in Andersen, ed.

Brown, Lester. 1978. *The global economic prospect: new sources of economic stress.* Worldwatch Paper no. 20. Washington, D.C.: Worldwatch Institute.

Brox, Ottar. 1964. Natural conditions, inheritance, and marriage in a north Norwegian fjord. *Folk* 6(1): 35–45.

Buchler, Ira, M. Fischer, and J. R. McGoodwin. 1986. Ecological structure, economics, and social organization: the Kapauku. Pp. 57–123 in De Meur, ed.

Buchler, Ira, and H. G. Nutini, eds. 1969. *Game theory in the behavioral sciences.* Pittsburgh, Pa.: University of Pittsburgh Press.

Burkenroad, Martin D. 1953. Theory and practice of marine fishery management. *Journal du Conseil International pour l'Exploration de la Mer* 18: 300–310.

Burrows, E. G., and M. E. Spiro. 1953. *An atoll culture: ethnography of Ifalik in the central Carolinas.* New Haven, Conn.: Human Relations Area Files.

Burton, Michael L., G. Mark Schoepfle, and Marc L. Miller. 1986. Natural resource anthropology. *Human Organization* 45: 261–69.

Butterworth, D. S. 1983. Assessment and management of pelagic stocks in the southern Benguela region. Unpublished ms. Department of Applied Mathematics, University of Cape Town.

Byron, Reginald. 1988. Luck and leadership: the management of decisions in Shetland fishing crews. *Mast* 1: 3–14.

Caddy, John F. 1984. Indirect approaches to regulation of fishing effort. FAO Fisheries Report no. 289, supp. 2: 63–75. Rome: FAO.

Carrier, James G. 1987. Marine tenure and conservation in Papua New Guinea: problems in interpretation. Pp. 142–67 in McCay and Acheson, eds.

Casteel, Richard W., and George I. Quimby, eds. 1975. *Maritime adaptations of the Pacific*. The Hague: Mouton.

Catarinussi, B. 1973. A sociological study of an Italian community of fishermen. Pp. 30–43 in Fricke, ed.

Chagnon, N., and R. Hames. 1979. Protein deficiency and tribal warfare in Amazonia: new data. *Science* 203: 910–13.

Chang, K. H. K. 1971. Institutional changes and the development of the fishing industry in a Japanese island community. *Human Organization* 30: 158–69.

Charest, P. 1979. Development of local and regional forms of political organization on the Gulf of St. Lawrence. Pp. 111–26 in Andersen, ed.

Charnov, E. 1973. Optimal foraging: some theoretical explorations. Ph.D. diss., University of Washington.

———. 1976a. Optimal foraging: the marginal value theorem. *Theoretical Population Biology* 9: 129–36.

———. 1976b. Optimal foraging: the attack strategy of a mantid. *American Naturalist* 109: 343–52.

Christy, Francis T., Jr. 1982. *Territorial use rights in marine fisheries: definitions and conditions*. FAO Fisheries Technical Paper no. 227. Rome: FAO.

Christy, Francis T., Jr., and Anthony Scott. 1965. *The common wealth in ocean fisheries*. Baltimore, Md.: Johns Hopkins University Press.

Clark, C. W. 1976. *Mathematical bioeconomics: the optimal management of renewable resources*. New York: Wiley and Sons.

———. 1977. Control theory in fisheries economics: frill or fundamental? Pp. 317–30 in L. Anderson, ed., *Economic impacts of extended fisheries jurisdiction*. Ann Arbor, Mich.: Ann Arbor Science Publishers.

Clark, C. W., and M. Mangel. 1977. Aggregation and fishing dynamics: a theoretical study of schooling and purse seine tuna. *Fisheries Bulletin of the U.S. National Marine Fisheries Service* 77(2): 471–80.

Clark, J. Desmond. 1970. *The prehistory of Africa*. New York: Praeger.

Clark, J. G. D. 1948. The development of fishing in prehistoric Europe. *Antiquaries Journal* 28: 45–85.

———. 1952. *Prehistoric Europe: the economic basis*. London: Methuen.

Clark, William C., and R. E. Munn, eds. 1986. *Sustainable development of the biosphere*. New York: Cambridge University Press.

Clepper, Henry, ed. 1979. *Marine recreational fisheries 4*. Washington, D.C.: Sportfishing Institute.

———. 1981. *Marine recreational fisheries 6*. Washington, D.C.: Sportfishing Institute.

Coe, Michael D., and Kent V. Flannery. 1967. *Early cultures and human ecology in south coastal Guatemala.* Smithsonian Contributions to Anthropology no. 3. Washington, D.C.: Smithsonian Press.

Cohen, J. 1975. Rural change in Ethiopia: the Chilalo Agricultural Development Unit. *Economic Development and Culture Change* 22(4): 580–614.

Cordell, John. 1973. Review of *The raft fishermen*, by Shepard Forman (1970). *American Anthropologist* 75: 1845–46.

———. 1974. The lunar tide fishing cycle in northeastern Brazil. *Ethnology* 13: 379–92.

———. 1978. Carrying capacity analysis of fixed-territorial fishing. *Ethnology* 17: 1–24.

———, ed. 1989. *A sea of small boats.* Cultural Survival Report no. 26. Cambridge, Mass.: Cultural Survival, Inc.

Craig, A. K. 1966. *Geography of fishing in British Honduras and adjacent coastal waters.* Louisiana State University Studies, Coastal Studies Series no. 14. Baton Rouge: Louisiana State University Press.

Creighton, Helen. 1950. *Folklore of Lunenberg County, Nova Scotia.* Ottawa: E. Cloutier, King's Printer.

Crutchfield, James A. 1979. Economic and social implications of the main policy alternatives for controlling fishing effort. *Journal of the Fisheries Research Board of Canada* 36: 742–52.

———, ed. 1959. *Biological and economic aspects of fisheries management.* Seattle: University of Washington.

Crutchfield, James A., and Giulio Pontecorvo. 1969. *The Pacific salmon fisheries.* Baltimore, Md.: Johns Hopkins University Press.

Dahl, C. 1988. Traditional marine tenure: a basis for artisanal fisheries management. *Marine Policy* 12: 40–48.

Danowski, F. 1980. *Fishermen's wives: coping with an extraordinary occupation.* University of Rhode Island Marine Bulletin no. 37. Kingston: University of Rhode Island.

Dasmann, R. F. 1974. Ecosystems. Paper presented at the Symposium on the Future of Traditional Primitive Societies, December, Cambridge, Eng.

Davenport, William H. 1960. Jamaican fishing: a game-theory analysis. Yale University Publications in Anthropology no. 59: 3–11. New Haven, Conn.: Yale University Press.

Davis, Anthony. 1984. Property rights and access management in the small boat fishery: a case study from southwest Nova Scotia. Pp. 133–64 in Lamson and Hanson, eds.

Davis, D. L. 1983. *Blood and nerves: an ethnographic focus on menopause.* Social and Economic Studies no. 28. St. John's: Institute of Social and Economic Research, Memorial University of Newfoundland.

Dawson, Chad P., and Bruce T. Wilkins. 1980. Social considerations associated with marine recreation fishing under the FCMA. *Marine Fisheries Review*, Dec.: 12–17.

DeGregori, T. R. 1974. Caveat emptor: a critique of the emerging paradigm of public choice. *Administration and Society* 6: 205–28.

De Meur, Giséle, ed. 1986. *New trends in mathematical anthropology*. London: Routledge and Kegan Paul.

Demsetz, H. 1967. Toward a theory of property rights. *American Economic Review* 62: 347–59.

Diamond, Norma. 1969. *K'un Shen: a Taiwan village*. New York: Holt, Rinehart, and Winston.

Dickie, L. M. 1969. The strategy of fishing. Unpublished ms. Dartmouth, Nova Scotia: Bedford Institute.

Dickinson, Joshua C., III. 1974. Fisheries of Lake Izabal, Guatemala. *Geographical Review* 64: 385–409.

Dorson, R. M. 1964. *Buying the wind*. Chicago: University of Chicago Press.

Doucet, Fernand J. 1984. Fishermen's quotas: one method of controlling fishing effort. FAO Fisheries Report no. 289, supp. 2: 161–65. Rome: FAO.

Douglas, Mary, and Aaron Wildavsky. 1983. *Risk and culture: an essay on the selection of technical and environmental dangers*. Berkeley: University of California Press.

Drucker, Philip. 1965. *Cultures of the North Pacific Coast*. San Francisco: Chandler.

Durrenberger, E. Paul. 1988. Shrimpers and turtles on the Gulf Coast: the formation of fisheries policy in the United States. *Maritime Anthropological Studies* 1: 196–214.

Durrenberger, E. Paul, and Gísli Pálsson. 1983. Riddles of herring and rhetorics of success. *Journal of Anthropological Research* 39: 323–35.

———. 1985. Peasants, entrepreneurs, and companies: the evolution of Icelandic fishing. *Ethnos* 50(1–2): 103–22.

———. 1986. Finding fish: the tactics of Icelandic skippers. *American Ethnologist* 13: 213–29.

———. 1987a. Ownership at sea: fishing territories and access to sea resources. *American Ethnologist* 14: 508–22.

———. 1987b. The grass roots and the state: resource management in Icelandic fishing. Pp. 370–92 in McCay and Acheson, eds.

———. 1988. Anthropology and fisheries management. *American Ethnologist* 15: 530–34.

Ebenreck, S. 1984. Stopping the raid on soil: ethical reflections on "sodbusting" legislation. *Agriculture and Human Values* 1(3): 3–9.

Economist, The. 1988. *The world in figures: editorial information compiled by The Economist*. Boston: G. K. Hall.

Emmerson, Donald R. 1980. *Rethinking artisanal fisheries development: Western concepts, Asian experiences*. World Bank Staff Working Paper no. 423. Washington, D.C.: World Bank.

FAO [Food and Agriculture Organization of the United Nations]. 1970. *FAO yearbook of fishery statistics*, vol. 31. Rome: FAO.

———. 1977. *FAO yearbook of fishery statistics*, vol. 45. Rome: FAO.

———. 1983. *Report of the expert consultation on the regulation of fishing effort (fishing mortality), January 17–26*. FAO Fisheries Report no. 289. Rome: FAO.

———. 1984. *Papers presented at the expert consultation on the regulation of fishing effort (fishing mortality), 17–26 January, 1983*. FAO Fisheries Report no. 289, supp. 2. Rome: FAO.

———. 1985. *Papers presented at the expert consultation on the regulation of fishing effort (fishing mortality), 17–26 January, 1983*. FAO Fisheries Report no. 289, supp. 3. Rome: FAO.

———. 1986. *FAO yearbook of fishery statistics*, vol. 63. Rome: FAO.

———. 1987. *FAO yearbook of fishery statistics*, vol. 65. Rome: FAO.

Faris, James C. 1966. *Cat Harbour: a Newfoundland fishing settlement*. St. John's: Institute of Social and Economic Research, Memorial University of Newfoundland.

Farmer, B. H. 1981. Review of *World systems of traditional resource management*, ed. Gary A. Klee (1980). *The Geographical Journal* 147: 238.

Ferdon, Edwin N., Jr. 1963. Polynesian origins. *Science* 141(3579): 499–505.

Firth, Raymond. 1946. *Malay fishermen: their peasant economy*. London: Routledge and Kegan Paul. 2d ed., 1966.

———. 1965. *Primitive Polynesian economy*. London: Routledge and Kegan Paul.

Forman, Shepard. 1967. Cognition and the catch: the location of fishing spots in a Brazilian coastal village. *Ethnology* 6: 417–26.

———. 1970. *The raft fishermen*. Bloomington: Indiana University Press.

Franke, R. W., and B. H. Chasin. 1980. *Seeds of famine: ecological destruction and the development dilemma in the West African Sahel*. Montclair, N.J.: Allenheld, Osmun.

Fraser, Thomas M., Jr. 1960. *Rusembilan: a Malay fishing village in southern Thailand*. Ithaca, N.Y.: Cornell University Press.

———. 1966. *Fishermen of South Thailand: the Malay villagers*. New York: Holt, Rinehart, and Winston.

Frazer, J. G. 1890. *The golden bough*. London: Macmillan.

Frick, Harold C. 1957. The optimum level of fisheries exploitation. *Journal of the Fisheries Research Board of Canada*. 14: 683–86.

Fricke, Peter H. 1985. Use of sociological data in the allocation of common property resources: a comparison of practices. *Marine Policy* 9: 39–52.

———. 1988. Memorandum prepared for the Fisheries Social Science Network, National Marine Fisheries Service, National Oceanic and Atmospheric Administration, U.S. Department of Commerce, Sept. 30.

———, ed. 1973. *Seafarer and community: towards a social understanding of seafaring*. London: Croom-Helm.

Furubotn, E. H., and S. Pejovich. 1972. Property rights and economic theory: a survey of recent literature. *Journal of Economic Literature* 10: 1137–62.

Fye, Paul M. 1977. Some implications of the Law of the Sea negotiations. Woods Hole Oceanographic Institution paper presented at the Textron annual management meeting, Feb. 26, Key Largo, Fla.

Gatewood, John B. 1983. Deciding where to fish: the skipper's dilemma in southeast Alaskan salmon seining. *Coastal Zone Management Journal* 10: 347–67.

———. 1984. Is the "skipper effect" really a false ideology? *American Ethnologist* 11: 378–79.

———. 1987. Information-sharing cliques and information networks. *American Ethnologist* 14: 777–78.

Gatewood, John B., and Bonnie J. McCay. 1988. Job satisfaction and the culture of fishing: a comparison of six New Jersey fisheries. *Maritime Anthropological Studies* 1(2): 103–28.

Georgianna, D. L., and W. V. Hogan. 1986. Production costs in Atlantic fresh fish processing. *Marine Resources Economics* 2(3): 275–92.

Gilles, J. L., and K. Jamtgaard. 1981. Overgrazing in pastoral areas: the commons reconsidered. *Sociologia Ruralis* 21 (Sept.): 129–41.

Gjessing, G. 1973. Maritime adaptations in north Norway's pre-history. Paper presented at the Ninth International Congress of Anthropological and Ethnological Sciences meetings, Chicago, Ill.

Glacken, C. J. 1955. *The great Loochoo.* Berkeley: University of California Press.

Gladwin, H. 1970. Decision making in the Cape Coast (Fante) fishing and fish marketing system. Ph.D. diss., Stanford University.

Gladwin, Thomas. 1970. *East is a big bird: navigation and logic on Puluwat Atoll.* Cambridge, Mass.: Harvard University Press.

Glantz, Michael H., and J. Dana Thompson. 1981. *Resource management and environmental uncertainty: lessons from coastal upwelling fisheries.* New York: John Wiley and Sons.

Gleick, James. 1987. *Chaos: making a new science.* New York: Penguin.

Godwin, R. K., and W. B. Shepard. 1979. Forcing squares, triangles and ellipses into a circular paradigm: the use of the commons dilemma in examining the allocation of common resources. *Western Political Quarterly* 32: 265–77.

González Casanova, Pablo. 1965. Internal colonialism and national development. *Studies in Comparative International Development* 1(4).

Goode, G. B. 1887. *The fisheries and fishing industries of the United States.* Washington, D.C.: Government Printing Office.

Goodlad, C. Alexander. 1972. Old and trusted, new and unknown: technological confrontation in the Shetland herring fishery. Pp. 61–81 in Andersen and Wadel, eds.

———. 1986. Regional fisheries management: the Shetland experience. Notes prepared for the Norwegian/Canadian Fisheries Management Workshop, Tromsø, Norway, June 16–21.

Gordon, H. Scott. 1953. An economic approach to the optimum utilization of fishery resources. *Journal of the Fisheries Research Board of Canada* 10: 442–57.

———. 1954. The economic theory of a common property resource: the fishery. *Journal of Political Economy* 62: 124–42.

———. 1957. Obstacles to agreement on control in the fishing industry. Pp. 65–72 in Turvey and Wiseman.

Gordon, William G. 1981. Marine recreational fisheries: outlook for the future. Pp. 9–12 in Clepper, ed.

Graham, M. 1935. Modern theory of exploiting a fishery, and application to North Sea trawling. *Journal du Conseil International pour l'Exploration de la Mer* 10: 264–74.

———. 1943. *The fish gate.* London: Faber and Faber. 2d ed., 1949.

Griffin, K. 1974. *The political economy of agrarian change: an essay on the Green Revolution.* Cambridge, Mass.: Harvard University Press.

Gross, D. 1975. Protein capture and cultural development in the Amazon basin. *American Anthropologist* 77: 526–49.

Grotius, Hugo. 1609. *Mare liberum* [*The freedom of the seas*]. Trans. R. V. D. Magoffin (1916). New York: Oxford University Press.

Gulland, John, ed. 1977. *Fish population dynamics.* Chichester: Wiley.

Hage, Per, and Frank Harary. 1983. *Structural models in anthropology.* Cambridge Studies in Social Anthropology. Cambridge: Cambridge University Press.

Hames, Raymond. 1987. Game conservation or efficient hunting? Pp. 92–107 in McCay and Acheson, eds.

———, ed. 1980. *Studies of hunting and fishing in the neotropics,* vol. 2 of *Working papers on South American Indians.* Bennington, Vt.: Bennington College.

Hamlisch, R., ed. 1962. *Economic effects of fishery regulation.* Report of an FAO expert meeting at Ottawa, 12–17 June, 1961. FAO Fisheries Reports no. 5. Rome: FAO.

Hardin, Garrett. 1968. The tragedy of the commons. *Science* 162: 1234–48.

———. 1988. Wilderness, a probe into "cultural carrying capacity." *Population and Environment* 10: 5–13.

Hardin, G., and J. Baden, eds. 1977. *Managing the commons.* San Francisco: W. H. Freeman.

Hardy, A. C. 1960. Was man more aquatic in the past? *The New Scientist* 7: 642–45.

Harris, Marvin. 1966. The cultural ecology of India's sacred cattle. *Current Anthropology* 7: 51–59.

———. 1980. *Culture, people, nature: an introduction to general anthropology,* 3d ed. New York: Harper and Row.

Harrison, Tom. 1970. *The Malays of south-west Sarawak before Malaysia*. East Lansing: Michigan State University Press.

Haviland, William A. 1983. *Human evolution and prehistory*, 2d ed. New York: Holt, Rinehart, and Winston.

Hay, J. 1959. *The run*. New York: Ballantine.

Heath, Anthony F. 1976. Decision making and transactional theory. Pp. 25–40 in Kapferer, ed.

Hennemuth, Richard C. 1979. Marine fisheries: food for the future. *Oceanus* 22(1): 2–12.

Herrington, R. 1962. Discussion comments, 73–74, 91–92, 95–96 in R. Hamlisch, ed.

Hewes, Gordon W. 1948. The rubric "fishing" and "fisheries." *American Anthropologist* 50: 238–46.

———. 1973. Indian fisheries productivity in pre-contact times in the Pacific salmon area. *Northwest Anthropological Research Notes* 7(2): 133–55.

Heyerdahl, Thor. 1963. Feasible ocean routes to and from the Americas in pre-Columbian times. *American Antiquity* 28: 482–88.

———. 1979. *Early man and the ocean: a search for the beginnings of navigation and seaborne civilizations*. Garden City, N.Y.: Doubleday.

Hilborn, Ray, and Wilf Luedke. 1987. Rationalizing the irrational: a case study in user group participation in Pacific salmon management. *Canadian Journal of Fisheries and Aquatic Sciences* 44: 1796–1805.

Holt, E. W. 1895. Present state of the Grimsby trawl fishery, with special reference to the destruction of immature fish. *Journal of the Marine Biological Association* 3: 339–446.

Homans, George Caspar. 1941. Anxiety and ritual: the theories of Malinowski and Radcliffe-Brown. *American Anthropologist* 43: 163–72.

Hubbs, Carl L., and Gunnar I. Roden. 1964. Oceanography and marine life along the Pacific Coast of Middle America. Pp. 143–86 in Robert C. West, vol. ed., and Robert Wauchope, gen. ed., *Natural environment and early cultures*, vol. 1 of *Handbook of Middle American Indians*. Austin: The University of Texas Press.

Hurst, James Willard. 1982. *Law and markets in United States history: different modes of bargaining among interests*. Madison: University of Wisconsin Press.

Idyll, C. P. 1978. *The sea against hunger*, updated ed. New York: Thomas Y. Crowell.

Jahoda, Gustav. 1969. *The psychology of superstition*. Baltimore, Md.: Penguin.

Jakobsson, Jakob. 1964. Recent developments in Icelandic herring purse seining. Pp. 294–305 in Hilmar Kristjónsson, ed., *Modern fishing gear of the world*, vol. 2. London: Fishing News.

James, M. 1959. Political and social limitations of fishery management. Pp. 23–30 in Crutchfield, ed.

Jennings, Jesse D., ed. 1983. *Ancient South Americans*. San Francisco: W. H. Freeman.

Jentoft, Svein. 1985. Models of fishery development: the cooperative approach. *Marine Policy* 9: 322–31.

———. N.d. Fisheries co-management: delegating government responsibility to fishermen's organizations. Unpublished ms.

Jentoft, Svein, and Trond Kristoffersen. 1987. Fishermen's self management: the case of the Lofoten Fishery. Paper presented at the Ninth International Seminar on Marginal Regions, July 5–11, Skye and Lewis, U.K.

———. 1989. Fishermen's co-management: the case of the Lofoten fishery. *Human Organization* 48: 355–65.

Jentoft, Svein, and Knut H. Mikalsen. 1987. Government subsidies in Norwegian fisheries: regional development or political favoritism? *Marine Policy* 11: 217–28.

Jett, Stephen C. 1983. Precolumbian transoceanic contacts. Pp. 337–93 in Jennings, ed.

Johannes, R. E. 1975. Exploitation and degradation of shallow marine food resources in Oceania. Pp. 47–71 in R. W. Force and B. Bishop, eds., *The impact of urban centers in the Pacific*. Honolulu, Hawaii: Pacific Science Association.

———. 1977. Traditional law of the sea in Micronesia. *Micronesia* 13(2): 121–27.

———. 1978. Traditional marine conservation methods in Oceania, and their demise. *Annual Review of Ecology and Systematics* 9: 349–64.

Johnson, A. 1982. Reductionism in cultural ecology: the Amazonian case. *Current Anthropology* 23: 413–28.

Johnson, Jeffrey C., and Duane Metzger. 1983. The shift from technical to expressive use of small harbors: the "play-full" harbors of southern California. *Coastal Zone Management Journal* 10: 429–41.

Johnson, Warren A. 1980. Europe. Pp. 165–88 in Klee, ed.

Jorion, Paul. 1976. To be a good fisherman you do not need any fish. *Cambridge Anthropology* 3(1): 1–12.

———. 1988. Going out or staying home: seasonal movements and migration strategies among Xwla and Anlo-Ewe fishermen. *Maritime Anthropological Studies* 1: 129–55.

Joseph, J. 1979. Highly migratory species: their conservation and management. Pp. 67–76 in Clepper, ed.

Kalland, Arne. 1984. Sea tenure in Tokugawa Japan: the case of Fukuoka Domain. Pp. 11–36 in Ruddle and Akimichi, eds.

Kapferer, B., ed. 1976. *Transaction and meaning: directions in the anthropology of exchange and symbolic behavior*. Philadelphia, Pa.: Institute for the Study of Human Issues.

Kassner, Jeffrey. 1988. The baymen of the Great South Bay, New York: a preliminary ecological profile. *Maritime Anthropological Studies* 1: 182–95.

Kearney, John. 1984. The transformation of the Bay of Fundy herring fisheries, 1976–1978: an experiment in fishermen-government co-management. Pp. 165–203 in Lamson and Hanson, eds.

Kemeney, John G., J. Laurie Snell, and Gerald L. Thompson. 1974. *Introduction to finite mathematics*, 3d ed. Englewood Cliffs, N.J.: Prentice-Hall.

Kesteven, G. L. 1976. Resources availability related to artisanal fisheries. Pp. 130–42 in Thomas S. Estes, ed., *Proceedings of the Seminar Workshop on Artisan Fisheries Development and Aquaculture in Central America and Panama*. Kingston: International Center for Marine Resources Development, University of Rhode Island.

Kirby Commission. 1983. *Navigating troubled waters: a new policy for the Atlantic fisheries*. Ottawa, Canada: Department of Fisheries and Oceans.

Kirch, Patrick V. 1979. Subsistence and ecology. Pp. 286–307 in Jesse D. Jennings, ed., *Prehistory of Polynesia*. Cambridge, Mass.: Harvard University Press.

Kitner, Kathi R., and John R. Maiolo. 1988. On becoming a billfisherman: a study of enculturation. *Human Organization* 47: 213–23.

Klee, Gary A., ed. 1980. *World systems of traditional resource management*. New York: John Wiley and Sons.

Kluckhohn, Clyde. 1942. Myths and rituals: a general theory. *Harvard Theological Review* 35: 44–79.

Knudson, K. E. 1970. Resource fluctuation, productivity and social organization on Micronesian coral islands. Ph.D. diss., University of Oregon.

Korten, D. C., ed. 1986. *Community management: Asian experience and perspectives*. West Hartford: Conn.: Kumarian Press.

Korten, D. C., and R. Klauss, eds. 1984. *People-centered development: contributions toward theory and planning framework*. West Hartford, Conn.: Kumarian Press.

Kottak, Conrad Phillip. 1966. The structure of equality in a Brazilian fishing community. Ph.D. diss., Columbia University.

Krebs, J. 1978. Optimal foraging: decision rules for predators. Pp. 23–63 in Krebs and Davies, eds.

Krebs, J., and N. Davies, eds. 1978. *Behavioural ecology: an evolutionary approach*. Oxford: Blackwell Scientific Publications.

Krech, S., ed. 1981. *Indians, animals, and the fur trade*. Athens: University of Georgia Press.

Lamson, Cynthia, and Arthur J. Hanson, eds. 1984. *Atlantic fisheries and coastal communities: fisheries decision-making case studies*. Halifax, Nova Scotia: Dalhousie Ocean Studies Programme, Dalhousie University.

Langdon, Steve J. 1984a. Alaskan native subsistence: current regulatory regimes and issues. Paper presented at the Roundtable Discussion of Subsistence, Oct. 10–13, Anchorage, Alaska.

———. 1984b. The perception of equity: social management of access in an Aleut fishing village. Paper presented at the annual meeting of the Society for Applied Anthropology, March 14–18, Toronto.

Lappé, Frances Moore, and Joseph Collins. 1977. *Food first: beyond the myth of scarcity*. Boston: Houghton Mifflin.

Lawson, R. 1984. *Economics of fisheries development*. New York: Praeger.

Leap, William L. 1977. Maritime subsistence in anthropological perspective: a statement of priorities. Pp. 251–63 in M. E. Smith, ed.

Leibhardt, Barbara. 1986. Among the bowheads: legal and cultural change on Alaska's North Slope coast to 1985. *Environmental Review* 10: 277–301.

Leschine, Thomas M. 1988. Ocean waste disposal management as a problem in decision-making. *Ocean and Shoreline Management* 11(1): 5–29.

Lessa, William A. 1966. *Ulithi: a Micronesian design for living*. New York: Holt, Rinehart, and Winston.

Lindblom, Charles E. 1959. The science of muddling through. *Public Administration Review* 19: 79–88.

———. 1979. Still muddling, not yet through. *Public Administration Review* 39: 517–26.

Lloyd, W. F. 1833. On the checks to population. Pp. 8–15 in Hardin and Baden, eds.

Löfgren, Orvar. 1972. Resource management and family firms: Swedish west coast fishermen. Pp. 82–103 in Andersen and Wadel, eds.

———. 1977. Fishermen's luck: magic and social tensions in two maritime communities. Mimeo. Lund, Sweden: Institute of Ethnology, University of Lund.

———. 1979. Marine ecotypes in preindustrial Sweden: a comparative discussion of Swedish peasant fishermen. Pp. 83–109 in Andersen, ed.

———. 1982. From peasant fishing to industrialized trawling: a comparative discussion of modernization processes in some North Atlantic regions. Pp. 151–76 in Maiolo and Orbach, eds.

Lummis, Trevor. 1983. Close-up: East Anglia. Pp. 182–223 in P. Thompson, T. Wailey, and T. Lummis.

———. 1985. *Occupation and society: the East Anglian fishermen, 1880–1914*. London: Cambridge University Press.

Luna, Julio. N.d. Artisanal fisheries: concepts for financing of development projects. Mimeo. Washington, D.C.: Inter-American Development Bank.

McCay, Bonnie J. 1976. Appropriate technology and coastal fishermen of Newfoundland. Ph.D. diss., Columbia University.

———. 1978. Systems ecology, people ecology, and the anthropology of fishing communities. *Human Ecology* 6: 397–442.

———. 1980. A fishermen's cooperative, limited: indigenous resource management in a complex society. *Anthropological Quarterly* 53: 29–38.

———. 1981a. Development issues in fisheries as agrarian systems. *Culture and Agriculture* 11 (May). Bulletin of the Anthropological Study Group on Agrarian Systems. Urbana-Champaign: University of Illinois.

———. 1981b. Optimal foragers or political actors? Ecological analyses of a New Jersey fishery. *American Ethnologist* 8: 356–82.

———. 1988. Muddling through the clam beds: cooperative management of New Jersey's hard clam spawner sanctuaries. *Journal of Shellfish Research* 7(2): 327–40.

McCay, Bonnie J., and James M. Acheson. 1987. Human ecology of the commons. Pp. 1–34 in McCay and Acheson, eds.

———, eds. 1987. *The question of the commons: the culture and ecology of communal resources*. Tucson: University of Arizona Press.

McDonald, D. 1977. Food taboos: a primitive environmental protection agency (South America). *Anthropos* 72: 734–48.

McEvoy, Arthur F. 1986. *The fisherman's problem: ecology and law in the California fisheries, 1850–1980*. Cambridge: Cambridge University Press.

McGoodwin, James R. 1976. Society, economy, and shark fishing crews in rural northwest Mexico. *Ethnology* 15: 377–91.

———. 1979. Pelagic shark fishing in rural Mexico: a context for co-operative action. *Ethnology* 18: 325–36.

———. 1980. The human cost of development. *Environment* 22(1): 25–31, 42.

———. 1984. Some examples of self-regulatory mechanisms in unmanaged fisheries. FAO Fisheries Report no. 289, supp. 2: 41–61. Rome: FAO.

———. 1986. The tourism-impact syndrome in developing coastal communities: a Mexican case. *Coastal Zone Management Journal* 14(1–2): 131–46.

———. 1987. Mexico's conflictual inshore Pacific fisheries: problem analysis and policy recommendations. *Human Organization* 46: 221–32.

McNabb, Steve. 1985. A final comment on measurement of the "skipper effect." *American Ethnologist* 12: 543–44.

Maiolo, John R. 1981. User conflicts in fisheries management. Pp. 81–92 in Clepper, ed.

Maiolo, John R., and Michael K. Orbach, eds. 1982. *Modernization and marine fisheries policy*. Ann Arbor, Mich.: Ann Arbor Science Publishers.

Maiolo, John R., and Paul Tschetter. 1982. Infrastructure investments in coastal communities: a neglected issue in studies of maritime adaptations. Pp. 203–24 in Maiolo and Orbach, eds.

Malinowski, Bronislaw. 1926. *Crime and custom in savage society*. London: Routledge and Kegan Paul.

———. 1954 (1948). *Magic, science, and religion*. New York: Doubleday Anchor Books.

Maréchal, Catherine, ed. 1988. *Women in artisanal fisheries*. IDAF Newsletter (June). Cotonou, Benin: FAO.

Martin, C. 1978. *Keepers of the game*. Berkeley: University of California Press.

Martin, Kent O. 1979. Play by the rules or don't play at all: space division and resource allocation in a rural Newfoundland fishing community. Pp. 277–98 in Andersen, ed.

Matsuda, Yoshiaki. 1972. Extension approach to the development of rural fishing villages on Hokkaido, Japan. Master's thesis, University of Georgia.

Meggars, Betty J. 1975. The transpacific origin of Mesoamerican civilization: a preliminary review of the evidence and its theoretical implications. *American Anthropologist* 77: 1–27.

Meggars, Betty J., and Clifford Evans. 1974. A transpacific contact in 3000 B.C. Pp. 97–104 in Zubrow, Fritz, and Fritz, eds.

Meighan, Clement W. 1959. The Little Harbor site, Catalina Island: an example of ecological interpretation in archaeology. *American Antiquity* 24: 383–405.

Meltzoff, Sarah Keene, and Edward Lipuma. 1986a. Hunting for tuna and cash in the Solomons: a rebirth of artisanal fishing in Malaita. *Human Organization* 45: 53–62.

———. 1986b. The troubled seas of Spanish fishermen: marine policy and the economy of change. *American Ethnologist* 13: 681–99.

MFCMA 1976. [Magnuson] fisheries conservation and management act. Public Law 94-265, 94th U.S. Congress, H.R. 200, Apr. 13, with amendments in 1981 and 1983.

Middleton, DeWight R. 1977. Changing economics in an Ecuadorian maritime community. Pp. 111–24 in M. E. Smith, ed.

Miller, Marc L. 1983. Culture, ethnography, and marine affairs. *Coastal Zone Management Journal* 10: 301–11.

Miller, Marc L., and John Van Maanen. 1979. "Boats don't fish, people do": some ethnographic notes on the federal management of fisheries in Gloucester. *Human Organization* 38: 377–85.

Mollett, Nina, ed. 1986. Fishery access control programs worldwide. *Proceedings of the Workshop on Management Options for the North Pacific Longline Fisheries, Orcas Island, Washington, April 21–25*. Alaska Sea Grant College Program, Report no. 86-4. Fairbanks.

Moore, Omar Khayyam. 1957. Divination: a new perspective. *American Anthropologist* 59: 69–74.

Moore, S. A., and H. S. Moore. 1903. *The history and law of fisheries*. London: Stevens and Haynes.

Morauta, L., J. Pernetta, and W. Heaney, eds. 1982. *Traditional conservation in Papua New Guinea: implications for today*. Monograph no. 16. Boroko, Papua New Guinea: Institute of Applied Social and Economic Research.

Morgan, Elaine. 1973. *The descent of woman*. New York: Bantam.

Morrill, Warren T. 1967. Ethnoicthyology of the Cha-Cha. *Ethnology* 6: 405–16.

Moseley, Michael Edward. 1975. *The maritime foundations of Andean civilization*. Menlo Park, Calif.: Cummings.

Mullen, P. B. 1969. The function of magic folk belief among Texas coastal fishermen. *Journal of American Folklore* 82: 214–25.

Murdock, G., and C. Provost. 1973. Factors in the division of labor by sex: a cross-cultural analysis. *Ethnology* 12: 203–25.

Nathan Associates. 1974. *The economic value of ocean resources to the United States.* Report submitted to the Congressional Research Service, Library of Congress, Washington, D.C.

National Research Council. 1986. *Proceedings of the Conference on Common Property Resource Management.* Washington, D.C.: National Academy Press.

Neal, Richard. 1982. Dilemma of the small-scale fishermen. *ICLARM Newsletter* 5(3): 7–9.

Nemec, Thomas F. 1972. I fish with my brother: the structure and behavior of agnatic-based fishing crews in a Newfoundland Irish outpost. Pp. 9–34 in Andersen and Wadel, eds.

Nietschmann, Bernard. 1974. When the turtle collapses, the world ends. *Natural History* 83(6): 34–43.

Nishimura, Asahitaro. 1973. *A preliminary report on current trends in marine anthropology.* Occasional Paper of the Center of Marine Ethnology no. 1. Tokyo: Waseda University.

———. 1975. Cultural and social change in the modes of ownership of stone tidal weirs. Pp. 77–88 in Casteel and Quimby, eds.

NMFS [U.S. National Marine Fisheries Service]. 1987. *NMFS work force by occupational series (permanent positions) FY 1987.* Washington, D.C.: NMFS.

Norbeck, Edward. 1954. *Takashima: a Japanese fishing village.* Salt Lake City: University of Utah Press.

Norr, James L., and Kathleen L. Norr. 1978. Work organization in modern fishing. *Human Organization* 37: 163–71.

Norr, Kathleen L. 1972. A south Indian fishing village in comparative perspective. Ph.D. diss., University of Michigan.

Orbach, Michael K. 1977. *Hunters, seamen, and entrepreneurs: the tuna seinermen of San Diego.* Berkeley: University of California Press.

———. 1983. The "success in failure" of the Vietnamese fishermen in Monterey Bay. *Coastal Zone Management Journal* 10(4): 331–46.

Orona, A. R. 1968. The social organization of the Margariteño fishermen, Venezuela. Ph.D. diss., University of California at Los Angeles.

Orth, Geoffrey C. 1987. Fishing in Alaska, and the sharing of information. *American Ethnologist* 14: 377–79.

Osamu, O., O. Mutsuo, T. Yoshiyuki, and Y. Junko. 1979. Fisheries in Asia: people, problems, and recommendations. International Foundation for Development Alternatives no. 14: 31–46.

Oto, Tokihiko. 1963. The taboos of fishermen. Pp. 107–21 in Richard M. Dorson, ed., *Studies of Japanese folklore.* Bloomington: Indiana University Press.

Paine, Robert. 1957. *Coast Lapp society,* vol. 1. Tromsø, Norway: Tromsø Museum.

Pálsson, Gísli. 1982. Territoriality among Icelandic fishermen. *Acta Sociologica* 25 (supp.): 5–13.

———. 1988. Models for fishing and models of success. *Mast* 1: 15–28.

Pálsson, Gísli, and E. Paul Durrenberger. 1982. To dream of fish: the causes of Icelandic skippers' fishing success. *Journal of Anthropological Research* 38(2): 227–42.

———. 1983. Icelandic foremen and skippers: the structure and evolution of a folk model. *American Ethnologist* 10: 511–28.

———. 1990. Systems of production and social discourse: The skipper effect revisited. *American Anthropologist* 92: 130–41.

Panayotou, Theodore. 1984. *Territorial use rights in fisheries.* FAO Fisheries Report no. 289, supp. 2: 153–60. Rome: FAO.

Peters, Pauline E. 1987. Embedded systems and rooted models: the grazing lands of Botswana and the commons debate. Pp. 171–94 in McCay and Acheson, eds.

Petersen, C. 1894. On the biology of our flatfishes and on the decrease of our flatfish fisheries. *Report of the Danish Biological Station* 4: 1–85.

Peterson, Susan B. 1973. Decisions in a market: a study of the Honolulu fish auction. Ph.D. diss., University of Hawaii at Manoa.

Peterson, Susan B., and Daniel Georgianna. 1988. New Bedford's fish auction: a study in auction method and market power. *Human Organization* 47: 235–41.

Petterson, John S. 1983. Policy and culture: the Bristol Bay case. *Coastal Zone Management Journal* 10(4): 313–30.

Pinkerton, Evelyn W. 1986. Co-cooperative management of local fisheries: a route to development. Paper presented at the Fisheries Co-Management Conference, May, Vancouver, British Columbia.

———. 1988. Co-operative management of local fisheries: a route to development. Pp. 257–273 in J. W. Bennett and J. R. Bowen, eds., *Production and autonomy: anthropological studies and critiques of development.* Lanham, Md.: Society for Economic Anthropology and University Press of America, Inc.

———, ed. 1989. *Co-operative management of local fisheries: new directions for improving management and community development.* Vancouver: University of British Columbia Press.

Pi-Sunyer, Oriol. 1976. The anatomy of conflict in a Catalan maritime community. Pp. 60–78 in J. B. Aceves, E. C. Hansen, and G. Levitas, eds., *Economic transformation and steady-state values: essays in the ethnography of Spain.* Flushing, New York: Queens College Publications in Anthropology.

———. 1977. Two states of technological change in a Catalan fishing community. Pp. 41–55 in M. E. Smith, ed.

Poggie, John J., Jr. 1978. Deferred gratification as an adaptive characteristic for small-scale fishermen. *Ethos*, Summer: 114–23.

————. 1979. Small-scale fishermen's beliefs about success and development: a Puerto Rican case. *Human Organization* 38: 6–11.

————. 1980a. Small-scale fishermen's psychocultural characteristics and cooperative formation. *Anthropological Quarterly* 53: 20–28.

————. 1980b. Ritual adaptation to risk and technological development in ocean fisheries: extrapolations from New England. *Anthropological Quarterly* 53: 122–29.

Poggie, John J., Jr., J. G. Bartee, and R. B. Pollnac. 1976. Multivariate correlates of success among small-scale fishermen in Puerto Rico. Paper presented at the Annual Meetings of the Northeastern Anthropological Association, March, Wesleyan University, Middletown, Conn.

Poggie, John J., Jr., and Carl Gersuny. 1974. *Fishermen of Galilee: the human ecology of a New England coastal community*. University of Rhode Island Marine Bulletin series no. 17. Kingston: University of Rhode Island.

Poggie, John J., Jr., and Richard B. Pollnac. 1988. Danger and rituals of avoidance among New England fishermen. *Mast* 1: 66–78.

Poggie, John J., Jr., R. B. Pollnac, and C. Gersuny. 1976. Risk as a basis for taboos among fishermen in southern New England. *Journal of the Scientific Study of Religion* 15(3): 257–62.

Polanyi, Karl. 1958. The economy as instituted process. Pp. 243–70 in Karl Polanyi, Conrad Arensberg, and Harry W. Pearson, eds., *Trade and markets in the early empires*. Glencoe, Ill.: The Free Press.

Pollnac, Richard B. 1976. Continuity and change in marine fishing communities. Anthropology Working Paper no. 10. Mimeo. Kingston: International Center for Marine Resource Development, University of Rhode Island.

————. 1982. Sociocultural aspects of technological and institutional change among small-scale fishermen. Pp. 225–47 in Maiolo and Orbach, eds.

————. 1984. Investigating territorial use rights among fishermen. Pp. 285–300 in Ruddle and Akimichi, eds.

————. 1987. People's participation in the small-scale fisheries development cycle. Anthropology Working Paper no. 47. Mimeo. Kingston: International Center for Marine Resource Development, University of Rhode Island.

————. 1988. Social and cultural characteristics of fishing peoples. *Marine Behavioral Physiology* 14: 23–39.

Pollnac, Richard B., and John J. Poggie, Jr. 1978. Economic gratification orientations among small-scale fishermen in Panama and Puerto Rico. *Human Organization* 37: 355–67.

————. 1979. The structure of job satisfaction among New England fishermen. Anthropology Working Paper no. 31. Mimeo. Kingston: International Center for Marine Resource Development, University of Rhode Island.

Pollnac, Richard B., and M. Robbins. 1972. Gratification patterns and modernization in rural Buganda. *Human Organization* 31: 63–72.

Pollnac, Richard B., and R. Ruiz-Stout. 1977. Small-scale fishermen's attitudes toward the occupation of fishing in the Republic of Panama. Pp. 16–20 in Richard B. Pollnac, ed., *Panamanian small-scale fishermen: society, culture, and change*. Marine Technology Report no. 44. Kingston: International Center for Marine Resources Development, University of Rhode Island.

Pontecorvo, Giulio. 1986. Opportunity, abundance, scarcity: an overview. Pp. 1–14 in Pontecorvo, ed.

———, ed. 1974. *Fisheries conflicts in the North Atlantic*. Cambridge, Mass.: Ballinger Press.

———. 1986. *The new order of the oceans: the advent of a managed environment*. New York: Columbia University Press.

Price, R. 1964. Magie et pêche à la Martinique. *L'Homme* 4: 84–113.

Prins, A. H. J. 1965. *Sailing from Lamu*. Assen, the Netherlands: Van Gorcum.

Pyke, G., R. Pulliam, and E. Charnov. 1976. Optimal foraging: a selective review of theories and tests. *Quarterly Review of Biology* 52: 137–54.

Radovich, J. 1975. Application of optimum sustainable yield theory to marine fisheries. Pp. 21–28 in Roedel, ed.

Reinman, F. M. 1967. Fishing: an aspect of Oceanic economy. *Fieldiana (Anthropology)* 56: 95–200.

Rettig, R. Bruce. 1987. Bioeconomic models: do they really help fishery managers? *Transactions of the American Fisheries Society* 116: 405–11.

Rockwood, C. E. 1973. A management program for the oyster resource of Apalachicola Bay, Florida. Unpublished report. Tallahassee: Florida State University.

Roedel, P., ed. 1975. *Optimum sustainable yield as a concept in fisheries management*. Washington, D.C.: American Fisheries Society.

Ross, E. 1978. Food taboos, diet, and hunting strategy: the adaptation to animals in Amazonian culture ecology. *Current Anthropology* 19: 1–36.

Ruddle, Kenneth, and Tomoya Akimichi. 1984. Introduction. Pp. 1–10 in Ruddle and Akimichi, eds.

———, eds. 1984. *Maritime institutions in the western Pacific*. Senri Ethnological Studies no. 17. Osaka: National Museum of Ethnology.

Ruddle, Kenneth, and R. E. Johannes, eds. 1985. *The traditional knowledge and management of coastal systems in Asia and the Pacific*. Jakarta: Regional Office for Science and Technology for Southeast Asia, United Nations Educational, Scientific, and Cultural Organization.

Russell, E. 1931. Some theoretical considerations on the overfishing problem. *Journal du Conseil International pour l'Exploration de la Mer* 6: 3–20.

Saetersdal, G. 1980. A review of past management of some pelagic stocks and its effectiveness. *Rapports et Procès-verbaux des Réunions, Conseil International pour l'Exploration de la Mer* 177: 505–12.

Sahlins, M. D. 1958. *Social stratification in Polynesia*. Seattle: University of Washington Press.

Schaefer, Milner B. 1954. Some aspects of the dynamics of populations important to the management of commercial marine fisheries. *Inter-American Tropical Tuna Bulletin* 1: 27–56.

———. 1957. Some considerations of the population dynamics and economics in relation to the management of the commercial marine fisheries. *Journal of the Fisheries Research Board of Canada* 14: 669–81.

———. 1959. Biological and economic aspects of the management of the commercial marine fisheries. *Transactions of the American Fisheries Society* 88: 100–104.

———. 1975. *Optimum sustainable yield as a concept in fisheries management*. Washington, D.C.: American Fisheries Society.

Schumacher, E. F. 1973. *Small is beautiful: economics as if people mattered*. New York: Harper and Row.

Schwerin, Karl H. 1970. *Winds across the Atlantic*. Mesoamerican Studies no. 6. Carbondale: Southern Illinois University.

Scott, A. D. 1955. The fishery: the objectives of sole ownership. *Journal of Political Economy* 63: 116–24.

Scott, P. 1951. Inshore fisheries of South Africa. *Economic Geography* 27(2): 123–47.

Scott, Stuart D., ed. 1967–73. *Archaeological reconnaissance and excavations in the Marismas Nacionales, Sinaloa and Nayarit, Mexico*. Buffalo, N.Y.: Department of Anthropology, State University of New York at Buffalo.

Service, Elman R. 1971. *Primitive social organization: an evolutionary perspective*. 2d ed. New York: Random House.

Shawcross, Wilfred. 1975. Some studies of the influences of prehistoric human predation on marine animal population dynamics. Pp. 39–66 in Casteel and Quimby, eds.

Shenkel, J. Richard. 1971. El Calón: revisited. Pp. 19–24 in S. D. Scott, ed.

Shoeffel, Penelope. 1985. Women in the fisheries of the South Pacific. Pp. 156–75 in R. V. Cole, ed., *Women in development in the South Pacific: barriers and opportunities*. Canberra: Development Studies Centre, Australian National University.

Simenstad, Charles A., James A. Estes, and Karl W. Kenyon. 1978. Aleuts, sea otters, and alternate stable-state communities. *Science* 200 (4340): 403–11.

Simoons, F. J., B. Schonfeld-Leber, and H. L. Issel. 1979. Cultural deterrents to use of fish as human food. *Oceanus* 22(1): 67–71.

Smith, Courtland L. 1988. Conservation and allocation decisions in fishery management. Pp. 131–38 in William J. McNeil, ed., *Salmon production, management, and allocation: biological, economic, and policy issues*. Corvallis: Oregon State University Press.

Smith, Homer W. 1953. *From fish to philosopher*. Boston: Little, Brown.

Smith, Leah J., and Susan B. Peterson, eds. 1982. *Aquaculture development in less developed countries: social, economic, and political problems*. Boulder, Colo.: Westview Press.

Smith, M. Estellie. 1982. Fisheries management: intended results and unintended consequences. Pp. 57–93 in Maiolo and Orbach, eds.

———. 1984. The triage of the commons. Paper presented at the Annual Meeting of the Society for Applied Anthropology, March 14–18, Toronto.

———. 1988. Fisheries risk in the modern context. *Mast* 1: 29–48.

———, ed. 1977. *Those who live from the sea: a study in maritime anthropology.* St. Paul, Minn.: West Publishing.

Smith, V. L. 1975. The primitive hunter culture: Pleistocene extinction and the rise of agriculture. *Journal of Political Economy* 83: 727–55.

Snedaker, Samuel C. 1971. The El Calón shell mound: an ecological anachronism. Pp. 15–18 in S. D. Scott, ed.

Spoehr, Alexander. 1980. *Protein from the sea: technological change in Philippine capture fisheries.* Ethnology Monographs no. 3. Pittsburgh, Pa.: Department of Anthropology, University of Pittsburgh.

———, ed. 1980. *Maritime adaptations: essays on contemporary fishing communities.* Pittsburgh, Pa.: University of Pittsburgh Press.

Stiles, R. Geoffrey. 1972. Fishermen, wives, and radios: aspects of communication in a Newfoundland fishing community. Pp. 35–60 in Andersen and Wadel, eds.

Stocks, Anthony. 1987. Resource management in an Amazon *varzea* lake ecosystem: the Cocamilla case. Pp. 108–120 in McCay and Acheson, eds.

Stoffle, Richard W., Danny L. Rasch, and Florence V. Jensen. 1983. Urban sports anglers and Lake Michigan fishery policies. *Coastal Zone Management Journal* 10(4): 407–27.

Stroud, Richard, ed. 1984. *Marine recreational fisheries 9.* Washington, D.C.: Sportfishing Institute.

Stuster, J. 1976. The material conditions of choice: an anthropological study of a California fishing community. Ph.D. diss., University of California at Santa Barbara.

———. 1978. Where "Mabel" may mean "sea bass." *Natural History* 87(9): 65–71.

Sudo, Ken-Ichi. 1984. Social organization and types of sea tenure in Micronesia. Pp. 203–30 in Ruddle and Akimichi, eds.

Suttles, W. P. 1968 (1960). Variation in habitat and culture on the Northwest Coast. Pp. 91–112 in Y. A. Cohen, ed., *Man in adaptation: the cultural present.* Chicago: Aldine.

———. 1974. *The economic life of the Coast Salish of Haro and Rosario Straits.* New York: Garland.

Swadling, P. 1976. Changes induced by human exploitation in prehistoric shellfish populations. *Mankind* 10: 156–62.

Taylor, R. 1963. The economic and social effect of having a limited number of licenses. *Fisheries Management Seminar*, pp. 4–11. Canberra: Department of Primary Industry.

Thompson, Paul, 1985. Women in the fishing: the roots of power between the sexes. *Comparative Studies in Society and History* 27: 3–32.

Thompson, Paul, Tony Wailey, and Trevor Lummis. 1983. *Living the fishing.* London: Routledge and Kegan Paul.

Thompson, Richard B. 1984. Marine recreational fisheries—update 1984. Pp. 15 –27 in Stroud, ed.

Thomson, David B. 1980. Conflict within the fishing industry. *ICLARM Newsletter* 3(3): 3–4.

Thorlindsson, Thorolfur. 1988. The skipper effect in the Icelandic herring fishery. *Human Organization* 47: 199–212.

Tiller, Per Olav. 1958. *Father absence and personality development of children in sailor families.* Nordisk Psykologi Monograph Series no. 9. Oslo: Munksgaard.

Tillion, Clement V. 1985. Fisheries management in Alaska. FAO Fisheries Report no. 289, supp. 3: 291–97. Rome: FAO.

Troadec, J. P., W. G. Clark, and J. A. Gulland. 1980. A review of some pelagic fish stocks in other areas. *Rapports et Procès-verbaux des Réunions, Conseil International pour l'Exploration de la Mer* 177: 252–77.

Turvey, R., and J. Wiseman, eds. 1957. *The economics of fisheries.* Rome: Food and Agriculture Organization of the United Nations.

U Thant. 1968. Cited in *International Planned Parenthood News* 168 (Feb.): 3.

Ulltang, Q. 1975. Catch per unit effort in the Norwegian purse seine fishery for Atlanto-Scandian (Norwegian spring spawning) herring. FAO Fisheries Technical Paper no. 155: 91–101. Rome: FAO.

———. 1980. Factors affecting the reaction of pelagic stocks to exploitation and requiring a new approach to assessment and management. *Rapports et Procès-verbaux des Réunions, Conseil International pour l'Exploration de la Mer* 177: 489 –504.

United Nations. 1982. Third United Nations Conference on Law of the Sea, Draft Convention of the Law of the Sea. U.N. Doc. A/Conf. 62/L-78/Rev. 3, Aug. 1981, released for signature Apr. 30, 1982.

Van Maanen, J., M. L. Miller, and J. C. Johnson. 1982. An occupation in transition: traditional and modern forms of commercial fishing. *Work and Occupations* 9: 193–216.

Vayda, Andrew P. 1988. Actions and consequences as objects of explanation in human ecology. *Environment, Technology, and Society* 51: 2–7.

Vickers, W. 1980. An analysis of Amazonian hunting yields as a function of settlement age. Pp. 7–29 in Hames, ed.

Von Neumann, John, and Oskar Morgenstern. 1944. *Theory of games and economic behavior.* Princeton, N.J.: Princeton University Press.

Wadel, Cato. 1972. Capitalization and ownership: the persistence of fishermen-ownership in the Norwegian herring fishery. Pp. 104–19 in Andersen and Wadel, eds.

Ward, Barbara. 1955. A Hong Kong fishing village. *Journal of Oriental Studies* 1: 195–214.

Warner, William W. 1977. The politics of fish: at sea with the international fishing fleet. *The Atlantic Monthly*, Aug.: 35–44.

———. 1983. *Distant water: the fate of the North Atlantic fisherman*. Boston: Little, Brown.

Watanabe, H. 1972. *The Ainu ecosystem*. Seattle: University of Washington Press.

Waugh, Geoffrey. 1984. *Fisheries management: theoretical developments and contemporary applications*. Boulder, Colo.: Westview Press.

White, D. J. 1954. *The New England fishing industry*. Cambridge, Mass.: Harvard University Press.

White, Douglas R. 1973. Mathematical anthropology. Pp. 369–446 in John J. Honigmann, ed., *Handbook of social and cultural anthropology*. Chicago: Rand McNally.

Williams, J. D. 1954. *The compleat strategyst*. Santa Monica, Calif.: RAND Corp.

Wilson, J. A. 1977. The tragedy of the commons—a test. Pp. 96–111 in Hardin and Baden, eds.

———. 1980. Adaptation to uncertainty and small numbers exchange: the New England fresh fish market. *Bulletin of the Journal of Economic Management Science* 11(2): 491–504.

Winterhalder, B. 1983. Opportunity-cost foraging models for stationary and mobile predators. *American Naturalist* 122: 73–84.

World Bank. 1987. *The World Bank Atlas, 1987*. Washington, D.C.: The World Bank.

Wulff, Robert M., and Shirley J. Fiske, eds. 1987. *Anthropological praxis: translating knowledge into action*. Boulder, Colo.: Westview Press.

Yesner, David R. 1977. Resource diversity and population stability among hunter-gatherers. *Western Canadian Journal of Anthropology* 7: 18–59.

———. 1980. Maritime hunter-gatherers: ecology and prehistory. *Current Anthropology* 26: 727–50.

Zengoryen. 1984. *Fisheries cooperative associations in Japan*. Tokyo: National Federation of Fisheries Cooperative Associations.

Zubrow, Ezra B. W., Margaret C. Fritz, and John M. Fritz, eds. 1974. *Readings from Scientific American, Archaeology: theoretical and cultural transformations*. San Francisco: W. H. Freeman.

Subject Index

Active indigenous regulation: control of rights and space, 123–26; political activism, 127–28; extralegal means, 128–31; biological control of stocks, 131–34; information management, 135–37; skill differences, 137–40; etiquette, 140–41; lessons of, for management, 141–42
Anthropology, maritime, *see* Maritime anthropology
Aquaculture, *see* Mariculture

Belief systems, 31–32, 138–40
Bioeconomic equilibrium, 71–72, 115–16. *See also* Commons
Bioeconomic modeling, 68–73
"Biosphere people" vs. "Ecosystem People," 41
Blue Revolution, 199

Catch: total annual fish, 1–2, 41, 97, 203; of small-scale fishers, 43. *See also* Factory ships; Fisheries management; Total allowable catch
Closed seasons or areas, 164–65
Colonialism, 99–100
Commons, theory of: development of, 89–91; weaknesses of, 92–96. *See also* Freedom of the seas doctrine
Community ownership, 125n
Competition, 18–19. *See also* Active indigenous regulation; Development and developmentalists; Problems in management
Count systems, 33
Crew recruitment and composition, 34
Cultures: defined, 21n–22n. *See also* Fishing cultures

Development and developmentalists: problems of, for small-scale fishers, 80–81; appropriate development, 113; technology and, 121; and problems in management, 160–61; future directions for, 199–201. *See also* Modernization
Drift-net fishing, 103

Economics and fisheries management, 73–77. *See also under* Passive indigenous regulation
Ecosystemic models, 87–88, 160
"Ecosystem People" vs. "Biosphere People," 41
Ecosystems, marine: fishers' knowledge of, 39–42; overfishing, 154
Environmental movement, 16–17
Exclusive economic zone (EEZ), 103–6; impact of, on small- and intermediate-scale fishers, 107

Factory ships, 101–3
Family relations, 24–25, 34, 35–39
Fisheries management: among early fishing peoples, 51–55; development of modern, 65–68; bioeconomic modeling and, 68–73; economics and, 73–77; policy formation and, 80–85; need for redeveloping, 86–88; humanizing, 108–11; of different types of fisheries, 152–55. *See also* Fisheries managers and fishers; Future directions in management; High-seas fisheries; Indigenous management; Problems in management; Strategies for management; Unregulated fisheries; *and under* Active indigenous regulation; Passive indigenous regulation

Library of Congress Cataloging-in-Publication Data

McGoodwin, James R.
    Crisis in the world's fisheries : people, problems, and
policies / James R. McGoodwin.
        p.   cm.
    Includes bibliographical references and index.
    ISBN 0-8047-1790-7 (cloth : acid-free paper) :
    I. Fishery management.   2. Fishery policy.   I. Title.
SH328.M39   1990
338.3'727—dc20   90-37484
                                                                CIP

♾ This book is printed on acid-free paper